# Minimal Submanifolds
## in Pseudo-Riemannian
## Geometry

# Minimal Submanifolds in Pseudo-Riemannian Geometry

HENRI ANCIAUX

*University of São Paulo, Brazil*

NEW JERSEY · LONDON · SINGAPORE · BEIJING · SHANGHAI · HONG KONG · TAIPEI · CHENNAI

*Published by*

World Scientific Publishing Co. Pte. Ltd.

5 Toh Tuck Link, Singapore 596224

*USA office:* 27 Warren Street, Suite 401-402, Hackensack, NJ 07601

*UK office:* 57 Shelton Street, Covent Garden, London WC2H 9HE

**British Library Cataloguing-in-Publication Data**
A catalogue record for this book is available from the British Library.

ISBN-13 978-981-4291-24-8
ISBN-10 981-4291-24-2

Printed in Singapore by World Scientific Printers.

To Marlene and Esteban

# Foreword

I met Henri Anciaux in 2000, at a time when I was reading an interesting paper where he solved partially a conjecture of Yong-Geun Oh. Since that time I have followed his scientific development and I have observed his skills to make his results more understandable by illustrating them with many examples.

In my opinion this book is a consequence of his particular vision of geometry. It also fills a long-standing gap, because many texts of pseudo-Riemannian geometry make use of physics to approach their topics, which sometimes implies that to learn basic concepts is a hard job for graduate students.

The introduction to pseudo-Riemannian and pseudo-Kähler geometries is enjoyable and easy to follow. The treatment made by the author about minimal, complex and Lagrangian submanifolds is clear and it will be useful for young researchers interested in these topics. The great quantity of examples not only helps to make clear the theory, but also allows an easier comprehension and a pleasant approach to the topic.

*F. Urbano*

# Preface

About 1755, the Turinese mathematician Lagrange derived the differential equation that a function satisfies when its graph minimizes the area among all surfaces with the same boundary. This achievement may be considered as the birth of the theory of minimal submanifolds, although Euler had discovered a few years before the first non-planar example of minimal surface, the catenoid. This is a surface of revolution (actually the only non-planar, minimal one) which owes its name to the fact that its generating curve is the catenary, the curve obtained by hanging freely a chain with uniform weight. The next step was taken by Meusnier who gave a geometric characterization of the minimal surface equation: the sum of the two principal curvatures of the surface vanishes at any point. He also re-discovered the catenoid[1] and did discover the helicoid, the surface made up by the trajectory of a straight line subject to a helical motion.

Since then this subject has been enjoying an enduring —although not at a constant rate— development until today. It has become an important one, with connections not only with the analysis of partial differential equations, but more surprisingly with complex analysis and even algebraic geometry, and has received contributions of major mathematicians, such as Poisson, Riemann, Weierstrass, Calabi... to name a few. While even some problems regarding the original setting, i.e. two-dimensional surfaces in Euclidean three-dimensional space, have proved hard to handle (to give an example, it was proved only in 2005 that the only non-planar minimal surface of Euclidean three-dimensional space which is embedded, complete and simply connected is the helicoid, see [Meeks, Rosenberg (2005)]), the theory has been generalized in several directions: instead of surfaces of Euclidean three-dimensional space, one can consider higher dimensional submanifolds,

---

[1] In those times free of "publish or perish" ideology, successive discoveries were frequent.

and replace the multi-dimensional Euclidean space by an arbitrary Riemannian manifold. Ultimately, one may observe that the assumption that the metric tensor is positive is unnecessary for most of the aspects of the topic, and that it can be dropped: the most general framework in which to address the study of minimal submanifolds is therefore that of pseudo-Riemannian geometry[2].

Although there is a huge literature on Riemannian geometry, and in particular on the theory of minimal submanifolds (without intending to be complete, we refer to [Osserman (1969)], [Chen (1973)], [Spivak (1979)], [Nitsche (1989)], [do Carmo (1992)], [Xin (2003)]), there are not many books about pseudo-Riemannian geometry, and those we know are focused on global analysis and/or physical applications rather than submanifold theory (see [O'Neill (1983)], [Kriele (1999)], [Palomo, Romero (2006)], [Alekseevsky, Baum (2008)]). The purpose of this book is twofold. We first give a basic introduction to the theory of minimal submanifolds, set from the beginning in the pseudo-Riemannian framework. This includes the important first variation formula, i.e. the generalization of Meusnier observation stating that a minimal submanifold has vanishing mean curvature vector. Our second aim is to present a selection of important results, ranging from classical ones, suitably generalized to the pseudo-Riemannian case (such as the Weierstrass representation, the classification of ruled minimal surfaces and the minimality of complex submanifolds) to more elaborate ones, including the classification of equivariant minimal hypersurfaces and a detailed study of Lagrangian submanifolds. It is hoped that this book, despite its imperfections, will be useful for graduate and postgraduate students, and researchers interested in this growing, exciting field.

The text is organized as follows: The first chapter provides a set of definitions and facts about pseudo-Riemannian geometry and submanifold theory, ending with the proof of the first variation formula. We only assume from the reader some knowledge of basic manifold theory (including the notion of vector fields, submanifolds, integration), but of course some acquaintance with Riemannian geometry, or at least with the classical theory of curves and surfaces, will ease the reading of the whole book. All the necessary material can be found, for example, in [Kühnel (2000)],

---

[2]Minimal surfaces have also been introduced in two close but different fields, namely in affine geometry and discrete geometry. Strictly speaking, these concepts are variants (interesting ones!) and not generalizations of the classical one discussed here. We refer the interested reader to [Simon (2000)] and [Bobenko, Schröder, Sullivan, Ziegler (2008)].

[do Carmo (1976)], [do Carmo (1992)]. The second chapter is devoted to the case of surfaces (two-dimensional submanifolds) in pseudo-Euclidean space. We first describe a variety of examples and give a first global result: the classification of ruled minimal surfaces. We also derive a generalized form of the classical Weierstrass representation formulae, a very important tool in the study of minimal surfaces. The third chapter is more technical: we introduce the simplest examples of non-flat pseudo-Riemannian manifolds, the *space forms,* and the notion of equivariant hypersurface. We classify minimal hypersurfaces of pseudo-Euclidean space and of space forms which are equivariant with respect to some natural group actions. The fourth chapter forgets for a while the subject of submanifolds and is devoted to the description of an important class of manifolds which enjoy a triple structure: pseudo-Riemannian, complex, and symplectic. Such manifolds are called *pseudo-Kähler manifolds* and generalize the concept of Kähler manifold. We describe some examples, such as the complex counter-parts of the *space forms,* and show that the tangent bundle of a pseudo-Kähler manifold is itself pseudo-Kähler. In the fifth chapter we come back to the core of the subject, focusing on two special classes of submanifolds appearing in pseudo-Kähler geometry, the complex and the Lagrangian ones. It is easily seen that a complex submanifold is always minimal, and the rest of the chapter is devoted to the study of Lagrangian submanifolds. In particular, equivariant minimal Lagrangian submanifolds of complex pseudo-Euclidean space and of complex space forms are classified, with the method already used in Chapter 3. The last chapter raises briefly the important question of whether a minimal submanifold, which is, by definition, a critical point of the volume, is actually an extremum of the volume functional, or not. We give both necessary and sufficient conditions for this to happen.

The notations used throughout the text should be transparent to the reader familiar with current mathematical textbooks. A word written *in italics* is being defined in the statement in which it appears, and the expression $A := B$ means that the mathematical quantity $A$ is defined to be equal to $B$. The symbol $\square$ marks the end of a proof.

Most of this book was written in Tralee, while I was a post doctoral fellow of the SFI (Science Foundation of Ireland). I had the opportunity to give two mini courses based on the material of this book. The first one, in January 2009, took place at the Technische Universität of Berlin, where I benefited an Elie Cartan Scholarship (Stiftung Luftbrückendank), while the second one was given in June 2010 at the Federal University of São Carlos, thanks to the support of the FAPESP (Fomento de Amparo à Pesquisa do Estado de São Paulo). I am grateful to both Mike Scherfner and Guillermo Lobos for taking care of everything in Berlin and São Carlos respectively. I warmly thank Kwong Lai Fun, from World Scientific Publishing, who has supported me along the process of preparing and editing the manuscript. I am also greatly indebted to Ildefonso Castro, Benoît Daniel, Brendan Guilfoyle and Pascal Romon, all of them both colleagues and friends, who carefully read earlier versions of this work and whose remarks reduced significantly, I hope, the number of the typos and imprecisions of the final text. I am honoured that Francisco Urbano kindly accepted to write the foreword of this book and it is my pleasure to thank him. This book is dedicated to my wife Marlene and my son Esteban.

*H. Anciaux*

# Contents

# Chapter 1

# Submanifolds in pseudo-Riemannian geometry

## 1.1 Pseudo-Riemannian manifolds

### 1.1.1 *Pseudo-Riemannian metrics*

A *pseudo-Riemannian* structure on a differentiable manifold $\mathcal{M}$ is simply a smooth, bilinear 2-form, called the *metric*, which is non-degenerate in the following sense: given a tangent vector $X$ at some point $x$,

$$\text{if } g(X,Y) = 0, \forall\, Y \in T_x\mathcal{M}, \text{ then } X = 0.$$

If in addition the metric satisfies $g(X,X) > 0$ for any non-vanishing tangent vector $X$, we say that the metric is *positive*, and that we have a *Riemannian* structure. Hence Pseudo-Riemannian geometry is simply a generalization of Riemannian geometry. A number of properties of positive metrics are no longer true in the general case, such as Cauchy-Schwartz inequality.

**Remark 1.** The non-degeneracy assumption implies the following important fact: given a tangent vector $X$ in $T_x\mathcal{M}$, if we know the value of $g(X,Y), \forall Y \in T_x\mathcal{M}$, then we can uniquely determine $X$.
In practice, we need only compute $g(X,X_i)$, where $(X_1, ..., X_m)$ is a basis of $T_x\mathcal{M}$: setting $g_{ij} = g(X_i, X_j), 1 \leq i, j \leq m$, the fact that the metric is non-degenerate implies that the matrix $[g_{ij}]_{1 \leq i,j \leq m}$ is invertible. Denoting the coefficients of the inverse matrix by $g^{ij}$ and writing $X = \sum_{i=1}^{m} \lambda_i X_i$, we find that $g(X, X_j) = \sum_{i=1}^{m} \lambda_i g(X_i, X_j) = \sum_{i=1}^{m} \lambda_i g_{ij}$. Multiplying by $g^{ij}$, we get $\lambda_i = \sum_{j=1}^{m} g^{ij} g(X, X_j)$, hence

$$X = \sum_{i,j=1}^{m} g^{ij} g(X, X_j) X_i. \tag{1.1}$$

A non-vanishing tangent vector $X$ will be called

- *positive* (or *spacelike*) if $g(X, X) > 0$;
- *negative* (or *timelike*) if $g(X, X) < 0$;
- *null* (or *lightlike*) if $g(X, X) = 0$.

These are the three possible *causal characters* of a vector. The terms in parenthesis come from relativity, the theory which first made use of pseudo-Riemannian geometry. More precisely the pseudo-Riemannian metric $\langle ., . \rangle_1 := -dx_1^2 + dx_2^2 + dx_3^2 + dx_4^2$ was introduced on the space $\mathbb{R}^4$, as a model of the space-time in special relativity[1]. In modern terms the pair $(\mathbb{R}^4, \langle ., . \rangle_1)$, sometimes written in abbreviated form $\mathbb{R}_1^4$ or $\mathbb{R}^{3,1}$, is called the Minkowski space (this space and its generalizations will be described in depth later on).

By Sylvester's theorem, at any point $x$ of $\mathcal{M}$, there exists an *orthonormal* basis $(e_1, ..., e_m)$ of $T_x\mathcal{M}$, in the sense that $g(e_i, e_j) = 0$ if $i \neq j$ and $|g(e_i, e_i)| = 1$ (we shall say that the $e_i$s are *unit* vectors). Moreover, the number $p$ of vectors of the basis which are negative (and hence the number $m - p$ of those which are positive) does not depend of the basis, nor on the point $x$. The pair $(p, m - p)$ is called the *signature* of $g$. For example, if the signature is $(0, m)$, the metric is Riemannian; if neither $p$ nor $m - p$ vanish, or equivalently if there exist null vectors, we say that the metric is *indefinite*. The metric of Minkowski space has signature $(1, 3)$. More generally, a pseudo-Riemannian manifold of signature $(1, m - 1)$ is referred to as a *Lorentzian* manifold. In the following, we shall set $\epsilon_i := g(e_i, e_i) = 1$ or $-1$ whenever we speak of an orthonormal basis $(e_1, ..., e_m)$. Formula (1.1) takes a much simpler form in the case of an orthonormal basis:

$$X = \sum_{i=1}^{m} \epsilon_i g(X, e_i) e_i. \tag{1.2}$$

The non-degeneracy assumption of the metric allows to define the important concept of the *trace* of a bilinear form with respect to $g$. Let $b$ be a bilinear form (possibly degenerate) on $T_x\mathcal{M}$, valued on any vector space $F$. Given a basis $(X_1, ..., X_m)$ of $T_x\mathcal{M}$, we claim that the quantity $\sum_{i,j=1}^{m} g^{ij} b(X_i, X_j) \in F$ depends only on $g$ and $b$, not on the choice of the basis: if $(Y_1, ..., Y_m)$ is another basis of $T\mathcal{M}$, there exist real constants $a_{ij}, 1 \leq i, j \leq n$ such that $Y_i = \sum_{k=1}^{m} a_{ik} X_k$. Thus we have $b(Y_i, Y_j) = \sum_{k,l=1}^{m} a_{ik} a_{jl} h(X_i, X_j)$. On the other hand, setting $\bar{g}_{ij} := g(Y_i, Y_j)$, we check that $\bar{g}^{ij} = a^{ik} a^{jl} g^{kl}$, where $[a^{ik}]$ is the inverse matrix of $[a_{ik}]$.

---

[1]Soon after, the theory of general relativity replaced this model by a more general pseudo-Riemannian manifold, thus triggering broad interest in the subject.

Therefore,

$$\sum_{i,j=1}^{m} \bar{g}^{ij} b(Y_i, Y_j) = \sum_{k,l=1}^{m} g^{kl} b(X_k, X_l).$$

Hence, the next definition makes sense:

**Definition 1.** The *trace* of a bilinear form $b$ with respect to $g$ is the quantity

$$\text{tr}(b) := \sum_{i,j=1}^{m} g^{ij} b(X_i, X_j).$$

**Remark 2.** Given an orthonormal basis $(e_1, ..., e_m)$, we have

$$\text{tr}(b) = \sum_{i=1}^{m} \epsilon_i b(e_i, e_i).$$

### 1.1.2 *Structures induced by the metric*

A metric is a very rich structure, in the sense that it induces several other structures in a canonical way. In the following we are going to review them quickly. We refer to [Kriele (1999)] or [O'Neill (1983)] for further details.

#### 1.1.2.1 *Volume*

A pseudo-Riemannian structure induces a volume structure, that is a $n$-density $dV$ defined by

$$dV(X_1, ..., X_m) = |\det([g(X_i, X_j)]_{1 \le i,j \le m})|^{1/2},$$

where $(X_1, ..., X_m)$ are $m$ tangent vectors to $\mathcal{M}$ at the point $x$. In particular, we may define the volume (possibly infinite) of the manifold $\mathcal{M}$, simply by integrating $dV$ on it:

$$\text{Vol}(\mathcal{M}) := \int_{\mathcal{M}} dV.$$

#### 1.1.2.2 *The Levi-Civita connection*

The differentiable structure of $\mathcal{M}$ allows to define the differentiation of a real function $f$ in the direction of a tangent vector $X \in T_x\mathcal{M}$, denoted by $X(f)(x)$ or $df_x(X)$: we set

$$X(f)(x) := \frac{d}{dt} f(\gamma(t)) \Big|_{t=0},$$

where $\gamma(t)$, $t \in I$ is a parametrized curve a curve $\gamma(t)$, such that $\gamma(0) = x$ and $\gamma'(0) = X$. To be rigorous, we should check that this definition does not depend on the choice of the curve $\gamma$ but only on the vector $X$, i.e. that if $\tilde{\gamma}(0) = \gamma(0)$ and $\tilde{\gamma}'(0) = \gamma'(0)$, then $\frac{d}{dt} f(\gamma(t))\big|_{t=0} = \frac{d}{dt} f(\tilde{\gamma}(t))\big|_{t=0}$. This easy task is left to the reader.

An *affine connection* $D$ or *covariant derivative* on $\mathcal{M}$ is, roughly speaking, a "way of differentiate a vector field" $Y$ of $\mathcal{M}$ along a parametrized curve $\gamma(t)$. The result is again a vector field defined along the curve. As in the case of the differentiation of real functions, this quantity does not actually depend on the curve $\gamma$, but rather on its velocity $\gamma'(t)$. We therefore denote it by $D_X Y$, where $X = \gamma'(t)$. Of course the expression $D_X Y$ makes sense if $X$ and $Y$ are two vector fields defined on an open subset of $\mathcal{M}$, the result being itself a vector field. We require furthermore that $D$ satisfies the two following rules:

$$D_X(fY) = f D_X Y + X(f)Y,$$
$$D_{fX}Y = f D_X Y,$$

where $f$ is a real function on $\mathcal{M}$.

A parametrized curve $\gamma(t)$ is said to be a *geodesic* with respect to the connection $D$ if $D_{\gamma'(t)}\gamma'(t) = 0, \forall t \in I$. Writing this equation in a local system of coordinates and using the theorem of existence for second order systems of ordinary differential equations, we get the local existence of geodesics: given a point $x \in \mathcal{M}$ and a tangent vector $X \in T_x\mathcal{M}$, there exists a real number $t_0$ and a unique geodesic $t \mapsto \gamma_{x,X}(t)$ defined on the interval $(-t_0, t_0)$, such that $\gamma_{x,X}(0) = x$ and $\gamma'_{x,X}(0) = X$.

There are many different affine connections on an arbitrary differentiable manifold, however the next result states that a pseudo-Riemannian structure comes with a canonical one:

**Theorem 1.** *There exists a unique affine connection $D$ on a pseudo-Riemannian manifold $(\mathcal{M}, g)$ satisfying*

(i) *$D$ has no torsion, i.e.*

$$D_X Y - D_Y X = [X, Y];$$

(ii) *$g$ is parallel with respect to $D$, i.e.*

$$Z\big(g(X, Y)\big) = g(D_Z X, Y) + g(X, D_Z Y).$$

*This unique connection is called* the Levi-Civita connection of $g$.

The proof of this result is based on a formula which is of interest on its own right:

**Lemma 1 (Koszul formula).** *Let $g$ be a metric and $D$ a connection satisfying conditions* (i) *and* (ii) *of the theorem above. Then, given a triple* $(X, Y, Z)$ *of vector fields, we have*

$$2g(D_X Y, Z) = X(g(Y, Z)) + Y(g(X, Z)) - Z(g(X, Y))$$
$$+ g([X, Y], Z) - g([X, Z], Y) - g([Y, Z], X).$$

*Proof.* Just take the right hand side term of the formula, use properties (i) and (ii) and do the obvious simplifications: one gets the left hand side term of the formula. $\square$

*Proof of Theorem 1.* We are going to give an explicit description of the Levi-Civita connection in terms of the metric $g$. By Remark 1, it is sufficient to define $g(D_X Y, Z)$, for any triple $(X, Y, Z)$ of vector fields. By Koszul formula we have an expression of $g(D_X Y, Z)$ depending only on the metric (and on the differentiable structure *via* the bracket). Hence it determines uniquely a connection $D$. We leave to the reader the tedious but straightforward task of checking that the structure $D$ defined in this way satisfies all the required properties of a connection without torsion and that it is parallel with respect to $g$. $\square$

**Example 1.** Let $(\mathbb{R}^m, \langle ., .\rangle_p)$ be the space of $m$-uples of real numbers, endowed with the pseudo-Riemannian metric of signature $(p, m-p)$ defined as follows: given two tangent vectors $X = (X_1, ..., X_m)$ and $Y = (Y_1, ..., Y_m)$, we set

$$\langle X, Y \rangle_p := \sum_{i=1}^{m} \epsilon_i X_i Y_i,$$

where $\epsilon_i = -1$ for $p$ of the $m$ indexes $i$, and $\epsilon_i = -1$ otherwise. We shall write for brevity $|X|_p^2 := \langle X, X \rangle_p$, although it may be a negative real number. A concise notation for $\langle ., .\rangle_p$ is the following:

$$\langle ., .\rangle_p := \sum_{i=1}^{m} \epsilon_i dx_i^2.$$

Most of the time (but not always), we shall set

$$\epsilon_i = -1 \text{ if } 1 \leq i \leq p \quad \text{and} \quad \epsilon_i = 1 \text{ if } p + 1 \leq i \leq m,$$

so that

$$\langle ., .\rangle_p = - \sum_{i=1}^{p} dx_i^2 + \sum_{i=p+1}^{m} dx_i^2.$$

In the following, the pair $(\mathbb{R}^m, \langle ., .\rangle_p)$ will be referred to as *pseudo-Euclidean space*.

It is easy to check that the Levi-Civita connection of $\langle ., .\rangle_p$ is the *flat connection* of $\mathbb{R}^m$, defined by

$$D_X Y(x) := dY_x(X),$$

i.e. the covariant derivative of the vector field $Y$ in the direction $X$ is simply the differential of $Y$ regarded as a map from $\mathbb{R}^m$ to $\mathbb{R}^m$, evaluated at $X$. Observe that *all* the metrics $\langle ., .\rangle_p, 0 \le p \le m$ have the *same* Levi-Civita connection, and in particular the same geodesics, which are the straight lines.

Denoting by $(x_1, ..., x_m)$ the rectangular coordinates of $\mathbb{R}^m$, the flat connection satisfies $\nabla_{\partial_{x_i}} \partial_{x_j} = 0, \forall i, j$. Although this is no longer true for an arbitrary connection and arbitrary coordinates, the next lemma shows that we can always find local coordinates in which the Levi-Civita connection of a pseudo-Riemannian manifold "looks flat", at least at a given point (such coordinates are called *normal coordinates*):

**Lemma 2.** *Let $(\mathcal{M}, g)$ be a pseudo-Riemannian manifold. Then given a point $x \in \mathcal{M}$ and an orthonormal basis $(e_1, ..., e_m)$ of $T_x\mathcal{M}$, there exist local coordinates $(x_1, ..., x_m)$ in a neighbourhood of $x$ such that $\partial_{x_i}(x) = e_i, \forall i$ and $\nabla_{\partial_{x_i}} \partial_{x_j}(x) = 0, \forall i, j$.*

*Proof.* The proof follows from the local existence of geodesics and from the following formula

$$\gamma_{x,X}(s) = \gamma_{x,tX}(1), \tag{1.3}$$

where $\gamma_{x,X}$ denotes the unique geodesic determined by the initial conditions $\gamma_{x,X}(0) = x$ and $\gamma'_{x,X}(0) = X$. We define the *exponential map* of $\mathcal{M}$ in $x$ by the expression $\exp_x(X) := \gamma_{x,X}(1)$ (so in particular, $\exp_x(0) = x$). Formula (1.3) and the smooth dependence of the geodesics with respect to initial conditions implies that the exponential map is well defined in a neighbourhood $U$ of $0$ in $T_x\mathcal{M}$ and that it is a diffeomorphism onto its image $U' := \exp_x(U)$. Since we are given an orthonormal basis $(e_1, ..., e_m)$ of $T_x\mathcal{M}$, $\mathbb{R}^m$ is itself diffeomorphic to $T_x\mathcal{M}$ by the map $(x_1, ..., x_m) \mapsto \sum_{i=1}^m x_i e_i$. The composition of these two diffeomorphisms defines therefore a system of coordinates $(x_1, ..., x_m)$ on $U'$. Moreover, Formula (1.3) shows that $d(\exp_x)_0 = Id$, which implies that $\partial_{x_i}(x) = d(\exp_x)_0(e_i) = e_i$. To conclude the proof, we use the fact that the coordinate curves $\gamma_i(t) :=$

$\exp(te_i)$ are geodesics, so that $0 = \nabla_{\gamma_i'(0)}\gamma_i'(0) = \nabla_{\partial_{x_i}}\partial_{x_i}(x)$. Analogously, for $i \neq j$, the "diagonal" curves $\gamma_{ij}(t) := \exp(t(e_i + e_j))$ being geodesics as well, we have

$$
\begin{aligned}
0 &= \nabla_{\gamma_{ij}'(0)}\gamma_{ij}'(0) \\
&= \nabla_{\partial_{x_i}}\partial_{x_i}(x) + 2\nabla_{\partial_{x_i}}\partial_{x_j}(x) + \nabla_{\partial_{x_j}}\partial_{x_j}(x) \\
&= 2\nabla_{\partial_{x_i}}\partial_{x_j}(x),
\end{aligned}
$$

which completes the proof. $\qquad\qquad\square$

### 1.1.2.3 *Curvature of a connection*

**Definition 2.** Let $D$ be any connection (not necessarily the Levi-Civita connection) on a manifold $\mathcal{M}$. The curvature tensor $R$ of $D$ is defined by

$$
R(X,Y)Z = D_Y D_X Z - D_X D_Y Z + D_{[X,Y]}Z.
$$

**Example 2.** Let $D$ be the flat connection on $\mathbb{R}^m$. Then a computation shows that $R(X,Y)Z = 0$, $\forall X, Y, Z$. This is why this connection is called *flat*.

In the case of the Levi-Civita connection of a metric, we can define some quantities related both to the curvature tensor and to the metric. The most important one is the following:

**Definition 3.** Let $P$ be a non-degenerate, two-dimensional linear subspace of the tangent space $T_x\mathcal{M}$. The *sectional curvature* of $P$ is defined to be

$$
K(P) = \frac{g(R(X,Y)X,Y)}{g(X,X)g(Y,Y) - g(X,Y)^2},
$$

where $X$ and $Y$ are two vectors spanning $P$ (this quantity does not depend on the choice of these vectors).

When the dimension of $\mathcal{M}$ is two, the unique sectional curvature that can be defined at $x$ is called the *Gaussian* curvature. In higher dimensions, there are other important notions of curvature obtained as traces of the tensor $R$: the Ricci curvature tensor and the scalar curvature. We refer to the literature for more details on this important subject, for example [Kriele (1999)], [do Carmo (1992)] (see also Exercises 1 of Chapter 3 and 6 of Chapter 4).

### 1.1.3   *Calculus on a pseudo-Riemannian manifold*

**Definition 4.** Let $(\mathcal{M}, g)$ be a pseudo-Riemannian manifold. Then

(i) the *gradient* $\nabla u$ of a smooth real function $u$ on $\mathcal{M}$ is the vector field defined by the identity (see Remark 1):

$$g(\nabla u, X) = du(X), \forall X \in T\mathcal{M};$$

(ii) the *divergence* $div X$ of a smooth vector field $X$ of $\mathcal{M}$ is the real map defined by:

$$div X = \text{tr}\big((X_1, X_2) \mapsto g(\nabla_{X_1} X, X_2)\big);$$

(iii) the *Laplacian* $\Delta u$ of a smooth real function $u$ on $\mathcal{M}$ is the divergence of its gradient:

$$\Delta u = div \nabla u.$$

A real function $u$ such that $\Delta u$ vanishes is said to be *harmonic*.

If $(e_1, ..., e_m)$ is an orthonormal frame on $(\mathcal{M}, g)$, we have the following expressions for the gradient and the divergence:

$$\nabla u = \sum_{i=1}^{m} \epsilon_i du(e_i) e_i,$$

$$div X = \sum_{i=1}^{m} \epsilon_i g(\nabla_{e_i} X, e_i).$$

The next theorem is a corollary of Stokes theorem, itself one of the most important results of geometry (see [Spivak (1965)]):

**Theorem 2 (Divergence theorem).** *If $X$ is a vector field on an oriented pseudo-Riemannian manifold $(\mathcal{M}, g)$ with boundary $\partial \mathcal{M}$, then*

$$\int_{\mathcal{M}} div X \, dV = \epsilon \int_{\partial \mathcal{M}} g(X, N) dV,$$

*where $N$ denotes the outward, unit normal vector of $\partial \mathcal{M}$ and $\epsilon := g(N, N)$. In particular, if $\partial \mathcal{M}$ is empty or if $X|_{\partial \mathcal{M}}$ vanishes, then*

$$\int_{\mathcal{M}} div X \, dV = 0.$$

*Proof.* Since $\mathcal{M}$ is oriented, there exists an $m$-form $\Theta$ such that $\int_\mathcal{M} dV = \int_\mathcal{M} \Theta$. Consider the $(m-1)$-form $\omega := X \lrcorner \Theta$. On the one hand, one computes that $d\omega = div X \, \Theta$. On the other hand, introduce a local, orthonormal, positively oriented frame $(e_1, ..., e_m)$ of an open subset of $\mathcal{M}$ such that $e_m|_{\partial\mathcal{M}} = N$. Then $(e_1, ..., e_{m-1})|_{\partial\mathcal{M}}$ is a local, orthonormal, positively oriented frame of $\partial\mathcal{M}$. Therefore, using Formula 1.2,

$$
\begin{aligned}
\omega(e_1, ..., e_{m-1})|_{\partial\mathcal{M}} &= \Theta(e_1, ..., e_{m-1}, X)|_{\partial\mathcal{M}} \\
&= \Theta\left(e_1, ..., e_{m-1}, \sum_{i=1}^m \epsilon_i g(X, e_i)\right)|_{\partial\mathcal{M}} \\
&= \epsilon g(X, N)\Theta(e_1, ..., e_m)|_{\partial\mathcal{M}}.
\end{aligned}
$$

Hence, by Stokes theorem,

$$
\int_\mathcal{M} div X \, \Theta = \int_\mathcal{M} d\omega = \int_{\partial\mathcal{M}} \omega = \epsilon \int_{\partial\mathcal{M}} g(X, N)\Theta. \qquad \square
$$

In particular, making $X = \nabla u$ in the divergence theorem, we obtain:

**Corollary 1.** *Let $u$ be a smooth function on a pseudo-Riemannian manifold $(\mathcal{M}, g)$. If $\partial\mathcal{M}$ is empty or if $\nabla u|_{\partial\mathcal{M}}$ vanishes, then $\int_\mathcal{M} \Delta u \, dV = 0$.*

Another important consequence of the divergence theorem is the following:

**Corollary 2.** *Let $u$ be a harmonic function on a compact, connected, Riemannian manifold $(\mathcal{M}, g)$. Then $u$ is constant.*

*Proof.* We just apply the divergence formula with $X = u\nabla u$. Since $div(u\nabla u) = u\Delta u + g(\nabla u, \nabla u)$, it implies that $\int_\mathcal{M} g(\nabla u, \nabla u)dV$ vanishes. Since the metric is positive, $\nabla u$ must vanish, so $u$ is constant. $\qquad \square$

## 1.2 Submanifolds

### 1.2.1 *The tangent and the normal spaces*

Submanifolds are higher dimensional analogs of curves. There are two dual ways of describing a submanifold $S$ of $\mathcal{M}$ of dimension $n$:

(i) As the *image* of an *embedding*, i.e. a differentiable map $f$ from a manifold $\tilde{S}$ of dimension $n$ into $\mathcal{M}$, satisfying a suitable *maximal rank* condition (we say that $f$ is an *immersion*) and such that $f : \tilde{S} \to S$ is a homeomorphism.

(ii) As the *pre-image* of a *submersion,* i.e. as the level set of a differentiable, $\mathbb{R}^{m-n}$-valued map on $\mathcal{M}$, also satisfying certain rank condition.

Both viewpoints, which are equivalent by the fundamental theorem of differential calculus, are useful. The first one proves that a submanifold enjoys the structure of a manifold. In particular, it has tangent vectors. The second one shows that these (abstract) tangent vectors can be identified with tangent vectors of $\mathcal{M}$. More precisely, at some point $x$ of $\mathcal{S}$, the tangent space $T_x\mathcal{S}$ can be identified with a linear subspace of $T_x\mathcal{M}$.

It is worth noting that the assumption that $f$ is a homeomorphism onto its image is not necessary for most of the definitions and theorems of the theory of submanifolds. This suggests to define an *immersed submanifold* $\mathcal{S}$ as being simply the image of a differentiable manifold $\tilde{\mathcal{S}}$ by an immersion $f$. In this case, there may exist two distinct points $x_1$ and $x_2$ of $\tilde{\mathcal{S}}$ with the same image $x = f(\tilde{x}_1) = f(\tilde{x}_2)$. In particular, the set of tangent vectors to $\mathcal{S}$ at $x$ is no longer a linear space if $df_{\tilde{x}_1} \neq df_{\tilde{x}_2}$. On the other hand, if $f$ is an immersion, every point $\tilde{x}$ of $\tilde{\mathcal{S}}$ enjoys a neighbourhood $U$ such that $f|_U$ is an embedding. In general, the image of an immersion may be a rather "bad" set of points (we shall see some examples later). We shall say that an immersion $f : \tilde{\mathcal{S}} \to \mathcal{M}$ is *proper* if the pre-image of a relatively compact open subset of $\mathcal{M}$ is a relatively compact subset of $\tilde{\mathcal{S}}$.

From the fact that $T_x\mathcal{S} \subset T_x\mathcal{M}, \forall x \in \mathcal{S}$, follows that we can define the restriction (denoted $g|_{\mathcal{S}}$ or simply $g$ if no confusion is possible) of $g$ to $\mathcal{S}$, traditionally called *the induced metric* (or *first fundamental form*). If $\mathcal{S}$ is defined as the image of an immersion $f : \tilde{\mathcal{S}} \to \mathcal{M}$, we can also pull back the metric $g$ on the abstract manifold $\tilde{\mathcal{S}}$. Geometrically this makes no difference and in particular $\text{Vol}(\mathcal{S}) = \text{Vol}(\tilde{\mathcal{S}})$, but the second viewpoint is sometimes useful technically.

The restriction of $g$ to $\mathcal{S}$ may be degenerate (for example, a curve whose velocity vectors are null). In the following, we shall almost always assume that it is not the case. In this case we shall say that $\mathcal{S}$ is a *non-degenerate* submanifold (some authors write *pseudo-Riemannian* submanifold). This assumption insures the pleasant fact that the *normal space* $N_x\mathcal{S}$ of $\mathcal{S}$ in $p$, defined to be the orthogonal of $T_x\mathcal{S}$, i.e.

$$N_x\mathcal{S} := (T_x\mathcal{S})^{\perp} = \{X \in T_x\mathcal{M}|\ g(X,Y) = 0, \forall Y \in T_x\mathcal{S}\},$$

makes a direct sum with $T_x\mathcal{S}$:

$$T_x\mathcal{M} = T_x\mathcal{S} \oplus N_x\mathcal{S}.$$

The dimension of $N_x\mathcal{S}$, i.e. the integer number $m - n$, is called the *co-dimension* of $\mathcal{S}$. Given a vector $X$ of $T_x\mathcal{M}$, we shall write the corresponding decomposition as $X = X^\top + X^\perp$ with $X^\top \in T_x\mathcal{S}$ and $X^\perp \in N_x\mathcal{S}$.

### 1.2.2 *Intrinsic and extrinsic structures of a submanifold*

Let $\mathcal{S}$ be a non-degenerate submanifold of $(\mathcal{M}, g)$. Recall that the restriction of $g$ to $\mathcal{S}$ induces a pseudo-Riemannian structure on $\mathcal{S}$. We have then *a priori* two natural connections on $\mathcal{S}$:

(i) The Levi-Civita connection of the induced metric $g|_\mathcal{S}$;

(ii) The "restriction" to $\mathcal{S}$ of $D$, i.e. given two tangent vector fields $X$ and $Y$, the tangential part of $D_X Y$. Again this is an intuitive statement whose proof is far from obvious: given two vector fields $X$ and $Y$ defined on $\mathcal{S}$, $D_X Y$ is actually meaningless; we must consider extensions $\bar{X}$ and $\bar{Y}$ of $X$ and $Y$ to $\mathcal{M}$ and show that the tangential part of $D_{\bar{X}}\bar{Y}\big|_\mathcal{S}$ does not depend on a particular choice of extension. We should also prove that the resulting object is actually a connection. We shall admit these facts and refer to the literature, for example [O'Neill (1983)], [Kriele (1999)].

It turns out that these two connections are the same one. In the following, we shall denote by $\nabla$ this connection. Of great interest is also the *normal part* of the restriction of $D$ to $\mathcal{S}$. It is called *second fundamental form* of $\mathcal{S}$ and denoted by $h$. We can summarize this proposition-and-definition by the so-called *Gauss formula*:

$$D_X Y = (D_X Y)^\top + (D_X Y)^\perp =: \nabla_X Y + h(X, Y).$$

**Proposition 1.** *$h$ is symmetric and tensorial, i.e. $h(X(x), Y(x))$ depends only on the values of $X$ and $Y$ at $x$.*

*Proof.* Let $X$ and $Y$ be two tangent vector fields, and $\bar{X}$ and $\bar{Y}$ be two extensions of them to $\mathcal{M}$. Then $[\bar{X}, \bar{Y}]|_\mathcal{S} = [X, Y]$ is tangent to $\mathcal{S}$. It follows that

$$0 = [\bar{X}, \bar{Y}]^\perp = (D_X Y)^\perp - (D_Y X)^\perp = h(X, Y) - h(Y, X),$$

so that $h$ is symmetric. We know that $(\nabla_X Y)(x)$, although it depends on the value of $Y$ on a *neighbourhood* of $x$, depends only on the value of $X$ *at the point $x$.* By the symmetry just proved, this implies that $h(X, Y)$ is tensorial. $\qquad \square$

We may also differentiate a normal vector field $\xi$ in the direction of a tangent vector and split the result into its tangential and normal parts (this is the *Weingarten formula*):

$$D_X\xi = (D_X\xi)^\top + (D_X\xi)^\perp := -A_\xi X + \nabla_X^\perp \xi.$$

The linear endomorphism of $X \mapsto A_\xi X$ is called *shape operator*.

**Proposition 2.** *The shape operator $A_\xi X$ is tensorial, i.e. it depends only on the values of $X$ and $\xi$ at $x$. It is moreover self-adjoint with respect to the induced metric, i.e. $g(A_\xi X, Y) = g(X, A_\xi Y)$. Finally, given $X, Y \in T_x S$ and $\xi \in N_x S$, we have the following relation between the shape operator and the second fundamental form:*

$$g(A_\xi X, Y) = g(h(X, Y), \xi). \tag{1.4}$$

*Proof.* We start by differentiating the identity $0 = g(Y, \xi)$ in the direction of the vector $X$:

$$0 = g(D_X Y, \xi) + g(Y, D_X \xi) = g(h(X, Y), \xi) + g(Y, -A_\xi X)$$

which gives Equation (1.4). Then the bilinearity, symmetry and tensoriality of the second fundamental form (Proposition 1) imply respectively the linearity, self-adjoint property and tensoriality of the shape operator $A_\xi$. $\square$

The next proposition, whose easy proof is ommited, describes the normal part of $D_X \xi$:

**Proposition 3.** *The operator $(D_X \xi)^\perp := \nabla_X^\perp \xi$ is a connection, called the normal connection of $S$, i.e. it is pointwise linear and satisfies the following property (Leibniz rule):*

$$\nabla_X^\perp(f\xi) = f\nabla_X^\perp \xi + X(f)\xi.$$

*Moreover, $g$ is parallel with respect to $\nabla^\perp$, i.e.*

$$X\big(g(\xi, \nu)\big) = g(\nabla_X^\perp \xi, \nu) + g(\xi, \nabla_X^\perp \nu).$$

We come back to the study of the shape operator. Taking into account the assumption that the induced metric is non-degenerate, Equation (1.4) shows that the second fundamental form and the shape operator carry exactly the same information. This is sometimes call the *extrinsic geometry* of $S$, in the sense that it depends not only on the induced metric on $S$ (which carries the *intrinsic geometry* of $S$), but really on how $S$ is "placed", or "embedded" in $\mathcal{M}$. In this regard, the word "shape operator" is quite evocative.

Given a normal vector $\xi$, a tangent vector $X$ is said to be a *principal direction* of $S$ with respect to $\xi$ if it is an eigenvector of the shape operator $A_\xi$, i.e. there exists a constant $\kappa$ such that $A_\xi X = \kappa X$. The constant $\kappa$ is then called a *principal curvature* of $S$. If the induced metric on $S$ is definite, it is a classical result of algebra that $A_\xi$ is diagonalisable with respect to $g$, i.e. there exists an orthonormal basis $(e_1, ..., e_n)$ of $T_x S$ of eigenvectors. Unfortunately, this pleasant fact no longer holds true if the induced metric is indefinite (see Exercise 4, Chapter 2).

**Definition 5.** A non-degenerate submanifold whose second fundamental form vanishes identically is called *totally geodesic*.

**Proposition 4.** *The totally geodesic submanifolds of the pseudo-Euclidean space $\mathbb{R}^m$ equipped with the flat metric $\langle ., . \rangle_p$ (see Example 1) are the open subsets of its (non-degenerate) linear subspaces. In particular, they are the same, independent of the signature $(p, m - p)$.*

*Proof.* Let $x$ denote a point of a totally geodesic submanifold $S$ of $(\mathbb{R}^m, \langle ., . \rangle_p)$ and $(e_1, ..., e_n)$ an orthonormal basis of $T_x S$. We also denote by $(N_1, ..., N_{m-n})$ an orthonormal basis of $N_x S$. Then there exists a neighbourhood $U$ of $x$ in $S$ which is a graph, i.e. there exists a map $F = (F^1, ..., F^{m-n})$ from an open subset $V$ of $\mathbb{R}^n$ into $\mathbb{R}^{m-n}$ such that $U$ is parametrized by

$$f(s) = f(s_1, ..., s_n) := \sum_{i=1}^{n} s_i e_i + \sum_{k=1}^{m-n} F^k(s) N_k.$$

It follows that the set of vector fields $X_i$ defined by

$$X_i := \frac{\partial f}{\partial s_i} = e_i + \sum_{k=1}^{m-n} \frac{\partial F^k}{\partial s_i} N_k$$

is a tangent frame to $S$. Also the set of vector fields $\xi_k$ defined by

$$\xi_k := -\sum_{i=1}^{n} \frac{\partial F^k}{\partial s_i} e_i + N_k$$

is a normal frame to $S$. We have seen in Example 1 that the covariant differentiation for the Levi-Civita connection $D$ of $\langle ., . \rangle_p$ amounts to the usual differentiation of $\mathbb{R}^m$. Hence

$$D_{X_i} X_j = \frac{\partial^2 f}{\partial s_i \partial s_j} = \sum_{k=1}^{m-n} \frac{\partial^2 F^k}{\partial s_i \partial s_j} N_k,$$

and

$$\langle h_{ij}, \xi_k \rangle_p = \langle (D_{X_i} X_j)^\perp, \xi_k \rangle_p = \frac{\partial^2 F^k}{\partial s_i \partial s_j}.$$

The submanifold $\mathcal{S}$ is totally geodesic if and only if $h_{ij}$ vanishes, $\forall i, j$, hence if and only if $\langle h_{ij}, \xi_k \rangle_p$ vanishes $\forall i, j, k$. We conclude that $\mathcal{S}$ is totally geodesic if and only if the maps $F^k$ are linear, so $\mathcal{S}$ is a linear subspace. $\square$

Rather than the vanishing of the whole second fundamental form, we shall be interested in the vanishing of some "part" of it, namely its trace. This leads us the most important definition of this book:

**Definition 6.** The trace of the second fundamental form $h$ with respect to the first fundamental form $g$ divided by $n$ is called the *mean curvature vector* of $\mathcal{S}$ and is denoted by $\vec{H}$.

It follows form the definition that, given an orthonormal basis $(e_1, ..., e_n)$ of $T_x\mathcal{S}$, the following formula holds:

$$\vec{H}(x) = \frac{1}{n} \sum_{i=1}^{n} \epsilon_i h(e_i, e_i).$$

**Corollary 3.** *The mean curvature vector of a totally geodesic submanifold vanishes identically.*

The following proposition shows that rescaling a metric has no effect in the set of minimal submanifolds of a manifold. This observation will be useful in Chapter 3.

**Proposition 5.** *Let $(\mathcal{M}, g)$ be a pseudo-Riemannian manifold. For real, non-vanishing $\lambda$, define the rescaled metric $g_\lambda := \lambda g$. Then the Levi-Civita connection of $g_\lambda$ is the same than that of $g$. Moreover the mean curvature vectors of a non-degenerate submanifold $\mathcal{S}$ with respect to $g$ and $g_\lambda$ respectively are given by*

$$\vec{H}^{g_\lambda} = \lambda^{-1} \vec{H}^g.$$

*Proof.* By the Koszul formula (Lemma 1), we see that the Levi-Civita connection of $g_\lambda$ is the same as the one of $g$. On the other hand, the decomposition $T\mathcal{M} = T\mathcal{S} \oplus N\mathcal{S}$ is the same for $g$ and $g^\lambda$ so the second fundamental form of a submanifold does not change. Then it is straightforward to check that the mean curvature vector changes by the rule $\vec{H}^{g_\lambda} = \lambda^{-1} \vec{H}^g$. $\square$

### 1.2.3  *One-dimensional submanifolds: Curves*

The local pseudo-Riemannian geometry of curves is so simple that its study does not deserve a whole chapter. On the other hand, curves will be frequently used as an auxiliary tool in the study of special higher dimensional submanifolds, so we find it convenient to state here the most basic facts of this regard.

### 1.2.3.1 *Arc length*

A parametrized curve $\gamma(t)$ from an interval $I$ into a pseudo-Riemannian manifold $(\mathcal{M}, g)$ is an immersion if and only if its velocity vector $d\gamma(\frac{d}{dt}) := \gamma'(t)$ does not vanish. We shall say that such a curve is *regular*. Moreover, the induced metric on $\gamma$ is non-degenerate if and only if $\gamma'(t)$ is non-null. We claim that if $\gamma'(t)$ is non-null, for all $t$ in $I$, we may reparametrize $\gamma$, i.e. find a diffeomorphism $s : t \mapsto s(t)$ such that $\left| g \left( \frac{d}{ds}\gamma(s), \frac{d}{ds}\gamma(s) \right) \right| = 1, \forall s \in I$. This new parameter $s$ is called the *arc length parameter*.

The proof of the existence of the arc length parameter goes as follows: given an arbitrary reparametrization $s(t)$, we have

$$\frac{d}{dt}\gamma = \frac{ds}{dt}\frac{d}{ds}\gamma,$$

so that

$$g\left( \frac{d}{dt}\gamma, \frac{d}{dt}\gamma \right) = \left( \frac{ds}{dt} \right)^2 g\left( \frac{d}{ds}\gamma(s), \frac{d}{ds}\gamma(s) \right).$$

The left hand side term does not vanish by the non-null assumption, hence we look for a function $s(t)$ such that

$$\frac{ds}{dt} = \left| g\left( \frac{d}{dt}\gamma, \frac{d}{dt}\gamma \right) \right|^{1/2},$$

in other words, the required $s(t)$ is simply a primitive of the right hand side of this expression.

### 1.2.3.2 *Curvature of a curve*

The second fundamental form of a curve reduces to a single normal vector field, sometimes called the *acceleration* of the curve:

$$h(\gamma', \gamma') = (D_{\gamma'}\gamma')^{\perp}.$$

For sake of simplicity, we write $\gamma'' = D_{\gamma'}\gamma'$. Assume that $\gamma$ is parametrized by the arc length. Differentiating the identity $g\left( \frac{d}{ds}\gamma, \frac{d}{ds}\gamma \right) = \pm 1$, we get that $\gamma''$ is normal to the curve, hence $h(\gamma', \gamma') = \gamma''$. Assume furthermore that $\gamma''$ is not null. Therefore there exists a positive function $\kappa$ and a unit vector field $\nu$ normal to the curve such that $\gamma'' = \kappa\nu$. Observe that $\kappa = |g(\gamma'', \gamma'')|^{1/2}$. The positive function $\kappa$ is called the *curvature* of the curve $\gamma$. In particular a curve with vanishing curvature is nothing but a geodesic.

### 1.2.3.3   *Curves in surfaces and the Frénet equations*

In a two-dimensional manifold (a surface), curves enjoy a slightly different notion of curvature: since the normal space to $\gamma$ at some point $\gamma(s)$ is one-dimensional, there exist exactly two unit normal vectors. Let us choose one of them and denote it by $\nu$. For example, if the surface is oriented, one may choose $\nu$ in such a way that the orthonormal frame $(\gamma', \nu)$ is positively oriented. Then the whole information concerning the extrinsic curvature of $\gamma$ is reduced to a scalar function, the shape operator $A_\nu$ being a linear endomorphism of a one-dimensional linear space. We define the (signed) curvature function of $\gamma$ with respect to $\nu$ to be the function $\kappa$ satisfying the identity $A_\nu \gamma' = \kappa \gamma'$. Observe that $\kappa$ may be negative and that replacing the unit vector $\nu$ by $-\nu$ results in replacing $\kappa$ by $-\kappa$. The notion of signed curvature, specific to dimension two, captures more information on the curve $\gamma$ than does the curvature defined in the section above. More precisely, setting $\tilde{\kappa} := |g(\gamma'', \gamma'')|^{1/2}$, the following expression holds: $|\kappa| = \tilde{\kappa}$. To see this, we differentiate the identity $g(\gamma', \nu) = 0$ with respect to the arc length parameter, getting that $g(\gamma'', \nu) = -g(\gamma', \nu') = \kappa g(\gamma', \gamma')$. Setting $\epsilon := g(\nu, \nu)g(\gamma', \gamma')$, it follows that $\gamma'' = \epsilon \kappa \nu$, so $|g(\gamma'', \gamma'')| = \kappa^2$, which implies the claimed expression. Observe furthermore that $\epsilon$ depends only on the signature of the metric on the surface, and not on the causal character of $\gamma$.

We have obtained also the *Frénet equations,* which we shall use several times along this book:
$$\begin{cases} \gamma'' = \epsilon \kappa \nu \\ \nu' = -\kappa \gamma'. \end{cases}$$
The fundamental theorem of existence of solutions of ordinary equations implies that given a real function $\kappa$ defined on some real interval $I$ containing 0 and initial conditions $(x, X)$, with $x \in \mathcal{M}$, $X \in T_x \mathcal{M}$, there exists a unique curve $\gamma$ defined on a subinterval $J$ of $I$ contained 0 such that $\gamma(0) = x$ and $\gamma'(0) = X$, and whose signed curvature is $\kappa$. In particular, making $\kappa = 0$, we recover the local existence of geodesics.

**Remark 3.** Some authors, for example [Kühnel (2000)], define the signed curvature to be the function defined by the identity[2] $\gamma'' = \kappa \nu$. Both definitions are equivalent when $\epsilon = 1$ (so in particular in the Riemannian setting), but give the opposite number when $\epsilon = -1$. One advantage of our choice is that it makes the curvature a principal curvature (the unique one, of course) of $\gamma$ with respect to $\nu$.

---

[2]In this case the Frénet equations become $\begin{cases} \gamma'' = \kappa \nu \\ \nu' = -\epsilon \kappa \gamma'. \end{cases}$

### 1.2.4   *Submanifolds of co-dimension one: Hypersurfaces*

When the co-dimension of a non-degenerate submanifold $S$ is one (in this case $S$ is said to be a *hypersurface*), the structure of the extrinsic curvature is simpler. Locally there always exists a unit normal vector field, i.e. such that $g(N, N) = \epsilon$, with $\epsilon = 1$ or $-1$ and then any normal vector $\xi$ is collinear to $N$. Since we have $\xi = \epsilon g(\xi, N)N$, all the information concerning the normal vector field $\xi$ is contained in the scalar function $g(\xi, N)$. Observe however that we first have to choose a unit vector field among the two possible ones.

For this reason, when dealing with a hypersurface, the second fundamental form is not defined to be a vector-valued tensor, but rather a scalar-valued one. Analogously, the mean curvature of a hypersurface refers usually to a scalar. However it is important to observe that both the second fundamental form and the mean curvature (recall that the latter is the trace of the former with respect to the first fundamental form) depend on the choice of a unit normal vector. The relations between the mean curvature vector $\vec{H}$ and the scalar mean curvature $H_N$ with respect to the unit vector $N$ are the following:

$$\vec{H} = \epsilon H_N N \qquad H_N = g(\vec{H}, N).$$

Moreover, the second fundamental form $h$ will be defined by

$$h(X, Y) := g(D_X Y, N) = -g(D_X N, Y),$$

and the Gauss equation becomes

$$D_X Y = \nabla_X Y + \epsilon h(X, Y)N.$$

Finally, differentiating the equation $g(N, N) = \epsilon$ shows that $g(D_X N, N)$ vanishes, hence the Weingarten formula simplifies to

$$D_X N = -A_N X.$$

**Definition 7.** A point $x$ of a hypersurface $S$ is said to be *umbilic* if the shape operator is a multiple of the identity, or equivalently the second fundamental form is proportional to the first one, i.e. there exists a constant $\kappa$ such that

$$h(X, Y) = \kappa g(X, Y).$$

In particular, $\kappa$ is a principal curvature of $S$ at $x$ and all tangent vectors are eigenvectors. A hypersurface is said to be *totally umbilic* if all its points are umbilic.

Of course if $\kappa$ vanishes, we fall back in the case of totally geodesic hypersurfaces. When the identity above holds, the mean curvature of $S$ is of course equal to $\kappa$. In Chapter 3 we shall characterize the totally umbilic hypersurfaces of some pseudo-Riemannian manifolds.

## 1.3 The variation formulae for the volume

### 1.3.1 *Variation of a submanifold*

We introduce now the concept of *variation* of a submanifold, which is, roughly speaking, a curve of submanifolds. More precisely, let $f : \tilde{S} \to \mathcal{M}$ be an immersion of an abstract manifold $\tilde{S}$. We consider a smooth map $F : \tilde{S} \times (-t_0, t_0) \to \mathcal{M}$ with $t_0 > 0$, such that $F(x, 0) = f(x)$. The vector field $X := \frac{\partial F}{\partial t}(x, 0)$ is called the *velocity vector* of the variation. We also make a technical requirement on $F$: we ask the variation (i) to *fix the boundary*, and (ii) to be *compactly supported*, i.e. there exists a relatively compact open subset $U$ of $\tilde{S}$ such that $U \cap \partial \tilde{S} = \emptyset$ and

$$F(x, t) = f(x), \ \forall x \in \tilde{S} \setminus U.$$

This assumption implies that $X$ is compactly supported, and that $X$ and all its derivatives vanish on $\partial \tilde{S}$.

The rank condition is an open condition, so the map $f_t : x \mapsto F(x, t)$ is an immersion for $t$ small enough, and thus $S_t := f_t(\tilde{S})$ is an immersed submanifold. Just as two different immersions may have the same image, two different maps $F$ may lead to the same variation $S_t$. However, the normal part $X^\perp$ of the vector field $X$ depends only on the variation $S_t$. Hence it will not be a surprise that it appears in the first variation formula (Theorem 4).

**Definition 8.** A non-degenerate submanifold $S$ of a pseudo-Riemannian manifold $(\mathcal{M}, g)$ is said to be *minimal* if its volume is critical with respect to any variation $S_t$, i.e.

$$\frac{d}{dt} \text{Vol}(S_t) \Big|_{t=0} = 0.$$

If $S$ is minimal, it is said to be *stable* if

$$\frac{d^2}{dt^2} \text{Vol}(S_t) \Big|_{t=0}$$

is monotone, i.e. positive or negative for all variation $S_t$.

We are following the tradition, coming from Lagrange, which imposed the word *minimal,* although it is particularly misleading: a minimal submanifold does not necessarily *minimize* the volume. The word *stationary,* much better, is sometimes used, and even the word *maximal* (because in certain situations the submanifold actually *maximizes* the volume). The thing to keep in mind is that for us to be minimal means to be critical for variations, i.e. the first order condition for a submanifold to be an extremum of volume functional. Similarly, the stability is nothing but the second order condition for a submanifold to be volume extremizing. We shall address in the last chapter the question of whether a minimal submanifold is an extremum (minimum or maximum) of the volume.

**Theorem 3.** *A compact submanifold $S$ of pseudo-Euclidean space $(\mathbb{R}^m, \langle ., . \rangle_p)$ is not minimal.*

*Proof.* Let $f : \tilde{S} \to \mathbb{R}^m$ a parametrization of $S$ and define

$$F : \tilde{S} \times (-t_0, t_0) \to \mathbb{R}^m$$
$$(x, t) \qquad \mapsto (1+t).f(x).$$

The compactness assumption insures that the variation $F$ satisfies the required conditions (i) and (ii). On the other hand an easy computation shows that $\mathrm{Vol}(S_t) = (1+t)^n \mathrm{Vol}(S)$, so that

$$\frac{d}{dt}\mathrm{Vol}(S_t)\Big|_{t=0} = n\mathrm{Vol}(S) \neq 0. \qquad \square$$

### 1.3.2 The first variation formula

**Theorem 4 (First variation formula).** *Let $S_t$ be a variation of a nondegenerate submanifold $S$ of a pseudo-Riemannian manifold $(\mathcal{M}, g)$. Then we have*

$$\frac{d}{dt}\mathrm{Vol}(S_t)\Big|_{t=0} = -\int_S g(n\vec{H}, X)dV,$$

*where $X$ is the velocity vector of the variation $S_t$.*

**Remark 4.** This formula enjoys an elegant interpretation in the language of functional analysis: assume for simplicity that $\tilde{S}$ is compact and has no boundary. Then the set $\mathcal{I}(\tilde{S}, \mathcal{M})$ of immersions of $\tilde{S}$ into $\mathcal{M}$ enjoys a (infinite dimensional) differentiable structure. A vector tangent to $\mathcal{I}(\tilde{S}, \mathcal{M})$ at $f$ may be identified with a vector field along $f(\tilde{S}) = S$, and $\mathcal{I}(\tilde{S}, \mathcal{M})$ is endowed with the natural metric $\tilde{g}$ defined by $\tilde{g}(X, Y) := \int_S g(X, Y)dV$. It turns out that the map $\mathrm{Vol} : \mathcal{I}(\tilde{\mathcal{S}}, \mathcal{M}) \to \mathbb{R}$ is smooth and the first variation formula simply says that the gradient of Vol with respect to $\tilde{g}$ is nothing but $-n\vec{H}$.

**Corollary 4.** *A submanifold $\mathcal{S}$ is minimal if and only if it has vanishing mean curvature vector.*

It is interesting to observe that this corollary provides a *pointwise* (so in particular easy-to-check) condition, i.e. the vanishing mean curvature, which is necessary for a *global* property to hold i.e. to minimize or to be critical point of the volume functional; moreover, it relates the field of *calculus of variations* (optimization of the volume) to the one of *differential geometry* (vanishing of a geometric quantity, the mean curvature vector).

We shall need the following lemma for the proof of the first variation formula:

**Lemma 3.** *Let $M(t)$, $t \in (-t_0, t_0)$ be a smooth curve of $n \times n$ real (or complex) invertible matrices. Then*

$$\frac{d}{dt} \det M(t) \Big|_{t=0} = \det M(0) \, \mathrm{tr}(M^{-1}(0) M'(0)).$$

*Proof.* Denoting by $c_1(t), ..., c_n(t)$ the columns of the matrix $M(t)$ and using the multi-linearity of the determinant, we compute

$$\frac{d}{dt} \det[c_1(t), ..., c_n(t)] \Big|_{t=0} = \sum_{i=1}^{n} \det[c_1(0), ..., c_i'(0), ..., c_n(0)].$$

Next, by the non-vanishing determinant assumption, the columns vectors $(c_1(0), ..., c_n(0))$ form a basis; we express the column vector $c_i'(0)$ in this basis:

$$c_i'(0) = \sum_{j=1}^{n} a_{ij} c_j(0)$$

and we deduce, using again the multilinearity of the determinant, that

$$\frac{d}{dt} \det M(t) \Big|_{t=0} = \sum_{i=1}^{n} a_{ii} \det[c_1(0), ..., c_i(0), ..., c_n(0)]$$

$$= \det M(0) \sum_{i=1}^{n} a_{ii} = \det M(0) \, \mathrm{tr}(M^{-1}(0) M'(0)). \qquad \square$$

*Proof of Theorem 4.* Let $\mathcal{S}_t$ a variation of $\mathcal{S}$ and $X$ its velocity vector field. By the assumption that $X$ is compactly supported, there exists $t_0$ such that $\forall t, |t| < t_0$, the map $f_t(x) := F(x,t)$ defines an immersion of $\tilde{S}$. In particular, given $t, |t| < t_0$, we have a induced metric on $\tilde{S}$, denoted $g_t$, and a volume form denoted $dV_t$. As two $n$-forms on a manifold of dimension $n$ are proportional, there exists a map $v(x,t)$ on $\tilde{S} \times (-t_0, t_0)$ such that $dV_t = v(x,t)dV$, and therefore

$$\frac{d}{dt}\text{Vol}(\mathcal{S}_t)\Big|_{t=0} = \int_{\tilde{S}} \frac{\partial v}{\partial t}(x,0)dV.$$

In order to calculate $v$, we introduce a local frame $(e_1,...,e_n)$ of $\tilde{S}$ and we ask that it is orthonormal for the metric $g_0$ (it is equivalent to say that $(df(e_1),...,df(e_n))$ is an orthonormal frame in $(\mathcal{M},g)$ tangent to $\mathcal{S}$). We then introduce the matrix

$$M(t) := [g_t(e_i,e_j)]_{1\leq i,j\leq n}$$

and claim that $v(x,t) = |\det M(t)|^{1/2}$. In fact

$$
\begin{aligned}
dV_t(e_1,...,e_n) &= |\det[g_t(e_i,e_j)]_{1\leq i,j\leq n}|^{1/2} \\
&= |\det M(t)|^{1/2} \\
&= |\det M(t)|^{1/2}dV(e_1,...,e_n).
\end{aligned}
$$

We are now in position to compute the first variation formula: by Lemma 3, and using the formula $\frac{d}{dt}|u|^{1/2} = \frac{u'u}{2|u|^{3/2}}$,

$$\frac{\partial v}{\partial t}(x,0) = \frac{1}{2}(\det M(0))^2 tr(M^{-1}(0)M'(0))|\det M(0)|^{-3/2}.$$

Since $(e_1,...,e_n)$ is orthonormal with respect to $g_0$, $M^{-1}(0) = M(0) = diag(\epsilon_1,...,\epsilon_n)$, so that

$$\frac{\partial v}{\partial t}(x,0) = \frac{1}{2}\sum_{i=1}^{n} \epsilon_i \frac{d}{dt}g_t(e_i,e_i)\Big|_{t=0}.$$

It remains to compute $\frac{d}{dt}g_t(e_i,e_i)\big|_{t=0}$. This is a bit involved in the general case of an arbitrary pseudo-Riemannian manifold, so we first give a proof in the simpler case where $(\mathcal{M},g)$ is the pseudo-Euclidean space $(\mathbb{R}^m, \langle.,.\rangle_p)$.

**Flat case.** From the Taylor expansion $f_t(x) = f(x) + tX(x) + o(t)$ we have, recalling the definition of the flat connection given in Example 1,

$$df_t(e_i) = df(e_i) + tdX(e_i) + o(t) = df(e_i) + tD_{e_i}X + o(t),$$

so that

$$g_t(e_i,e_i) = g(df_t(e_i),df_t(e_i)) = g(df(e_i),df(e_i)) + 2g(df(e_i),D_{e_i}X) + o(t),$$

and thus

$$\frac{d}{dt}g_t(e_i, e_i)\Big|_{t=0} = 2g(e_i, D_{e_i}(X^\top + X^\perp))$$

$$= 2g(e_i, D_{e_i}X^\top) + 2g(e_i, D_{e_i}X^\perp)$$

$$= 2g(e_i, \nabla_{e_i}X^\top) + 2g(e_i, -A_{X^\perp}e_i).$$

Using Equation (1.4) of Section 1.2.2, it follows that

$$\frac{\partial v}{\partial t}(x, 0) = \sum_{i=1}^n \epsilon_i \Big(g(e_i, \nabla_{e_i}X^\top) - g(X^\perp, h(e_i, e_i))\Big)$$

$$= div(X^\top) - g(X^\perp, n\vec{H}).$$

To conclude, we use the divergence theorem (Theorem 2) and the fact that $X$ vanishes on $\partial\mathcal{S}$, to get:

$$\frac{d}{dt}\text{Vol}(\mathcal{S}_t)\Big|_{t=0} = -\int_{\mathcal{S}} g(X^\perp, n\vec{H})dV = -\int_{\mathcal{S}} g(X, n\vec{H})dV.$$

**General case.** The idea of the proof is to use $F$ as an immersion when possible, and to check that when it is not possible, the variation of the volume is zero. We thus define the two open subsets $U_1 := \{x \in \tilde{\mathcal{S}}, X^\perp(x) \neq 0\}$ and $U_2 := int\{x \in \tilde{\mathcal{S}}, X^\perp(x) = 0\}$. We thus have

$$\frac{d}{dt}\text{Vol}(\mathcal{S}_t) = \frac{d}{dt}\text{Vol}(f_t(U_1)) + \frac{d}{dt}\text{Vol}(f_t(U_2)).$$

We first observe that the assumption $X^\perp \neq 0$ is equivalent to the fact that $F$ is an immersion for $t$ small enough, say for $t \in (-t_1, t_1) \subset (-t_0, t_0)$. Therefore we can pull back the metric $g$ and the vector fields $e_1, ..., e_n$, and $X$ to $U_1 \times (-t_1, t_1)$. Moreover, using the fact that $[X, e_i] = 0$ at $t = 0$, we have

$$\bar{\nabla}_X e_i = \bar{\nabla}_{e_i} X,$$

where we denote by $\bar{\nabla}$ the Levi-Civita connection of $U_1 \times (-t_1, t_1)$. It follows that

$$\frac{d}{dt}g(e_i, e_i)\Big|_{t=0} = 2g(\bar{\nabla}_X e_i, e_i) = 2g(\bar{\nabla}_{e_i} X, e_i)$$

$$= 2g(\bar{\nabla}_{e_i}X^\top, e_i) + 2g(\bar{\nabla}_{e_i}X^\perp, e_i)$$

$$= 2g(\nabla_{e_i}X^\top, e_i) + 2g(-A_{X^\perp}e_i, e_i).$$

At this stage we proceed exactly as in the flat case and we conclude that

$$\frac{d}{dt}\text{Vol}(f_t(U_1))\Big|_{t=0} = -\int_{U_1} g(X^\perp, n\vec{H})dV = -\int_{U_1} g(X, n\vec{H})dV.$$

It remains to be seen that on $U_2$, the variation of the volume is zero at the first order. The reason for this is that, as $X^\perp$ vanishes, the variation is tangential and $f_t$ is, up to first order, a diffeomorphism of the submanifold $U_2$ itself. Therefore its global volume remains constant (up to first order again!). Hence

$$\frac{d}{dt}\text{Vol}(f_t(U_2))\bigg|_{t=0} = 0,$$

and the proof is complete. $\qquad\qquad\qquad\qquad\qquad\qquad\qquad\qquad\qquad$ □

### 1.3.3 *The second variation formula*

**Theorem 5 (Second variation formula).** *Let $\mathcal{S}_t$ be a normal variation of a minimal submanifold $S$ of a pseudo-Riemannian manifold $(\mathcal{M}, g)$. Then we have:*

$$\frac{d^2}{dt^2}\text{Vol}(\mathcal{S}_t)\bigg|_{t=0} = \int_S \left( g(\nabla^\perp X, \nabla^\perp X) - g(A_X, A_X) + g(R^\perp(X), X) \right) dV,$$

*where*

$$g(\nabla^\perp X, \nabla^\perp X) := \text{tr}\left( (Y_1, Y_2) \mapsto g(\nabla_{Y_1} X, \nabla_{Y_2} X) \right),$$

$$g(A_X, A_X) := \text{tr}\left( (Y_1, Y_2) \mapsto g(A_X Y_1, A_X Y_2) \right),$$

*and*

$$R^\perp(X) := \text{tr}\left( (Y_1, Y_2) \mapsto R(Y_1, X)Y_2 \right).$$

**Remark 5.** Given an orthonormal frame $(e_1, ..., e_n)$, we have the following expressions for the quantities appearing in the formula above:

$$g(\nabla^\perp X, \nabla^\perp X) = \sum_{i=1}^n \epsilon_i g(\nabla^\perp_{e_i} X, \nabla^\perp_{e_i} X),$$

$$g(A_X, A_X) = \sum_{i=1}^n \epsilon_i g(A_X e_i, A_X e_i),$$

and

$$R^\perp(X) = \sum_{i=1}^n \epsilon_i R(e_i, X) e_i.$$

*Proof of Theorem 5.* We have seen in the proof of Theorem 4 that there is no variation of the volume on the set $int\{x \in \tilde{S}, X^\perp(x) = 0\}$. We may therefore assume without loss of generality that $X^\perp(x) \neq 0$. In particular, $F : \tilde{S} \times (-t_1, t_1)$ is an immersion.

**Lemma 4.** *We have*

$$\left.\frac{dg^{ij}}{dt}\right|_{t=0} = 2\epsilon_i\epsilon_j g(h(e_i, e_j), X),$$

*where* $[g^{ij}]_{1 \leq i,j \leq n}$ *denotes the inverse matrix of* $[g_{ij}]_{1 \leq i,j \leq n} :=$ $[g(e_i, e_j)]_{1 \leq i,j \leq n}$.

*Proof.* We first calculate $\left.\frac{dg_{ij}}{dt}\right|_{t=0}$. For this purpose, recall that $[e_i, X] = 0$ on $\tilde{S} \times (-t_1, t_1)$, so that

$$
\begin{aligned}
\left.\frac{d}{dt}g(e_i, e_j)\right|_{t=0} &= g(\bar{\nabla}_X e_i, e_j) + g(e_i, \bar{\nabla}_X e_j) \\
&= g(\bar{\nabla}_{e_i} X, e_j) + g(e_i, \bar{\nabla}_{e_j} X) \\
&= -g(A_X e_i, e_j) - g(e_i, A_X e_j) \\
&= -2g(h(e_i, e_j), X).
\end{aligned}
$$

To complete the proof of the lemma, we differentiate the identity $\sum_{k=1}^n g^{ik} g_{kj} = \delta_{ij}$ with respect to $t$, getting:

$$\sum_{k=1}^n \left(\frac{dg^{ik}}{dt} g_{kj} + g^{ik}\frac{dg_{kj}}{dt}\right) = 0,$$

and evaluate the resulting expression at $t = 0$:

$$\left.\frac{dg^{ij}}{dt}\right|_{t=0} \epsilon_j + \epsilon_i \left.\frac{dg_{ij}}{dt}\right|_{t=0} = 0.$$

Finally we obtain the claimed formula:

$$\left.\frac{dg^{ij}}{dt}\right|_{t=0} = -\epsilon_i\epsilon_j \left.\frac{dg_{ij}}{dt}\right|_{t=0} = 2\epsilon_i\epsilon_j g(h(e_i, e_j), X). \qquad \square$$

We now prove the second variation formula. Differentiating the first variation formula

$$\frac{d}{dt}\mathrm{Vol}(S_t) = -\int_{\tilde{S}} g(n\vec{H}_t, X_t)dV_t,$$

we get

$$\frac{d^2}{dt^2}\mathrm{Vol}(\mathcal{S}_t)\bigg|_{t=0} = -\int_{\tilde{S}} \frac{d}{dt} g(n\vec{H}_t, X_t) dV_t\bigg|_{t=0}$$

$$= -\int_{\tilde{S}} \frac{d}{dt} g(n\vec{H}_t, X_t)\bigg|_{t=0} dV_0 - \int_{\tilde{S}} g(n\vec{H}_0, X_0) \frac{d}{dt} dV_t\bigg|_{t=0}.$$

Since $\vec{H}_0$ vanishes identically, the second term of the last the expression above vanishes and it remains to calculate $\frac{d}{dt} g(n\vec{H}, X)$ at $t = 0$. We are working pointwise, so we may use Lemma 2 of Section 1.1.2 and assume the existence of local coordinates $(x_1, ..., x_n)$ in a neighbourdhood of a point $x$ such that $\partial_{x_i}(x) = e_i(x)$ and $\nabla_{e_i} e_j(x) = 0$. It follows from the Gauss formula that $D_{e_i} e_j = h(e_i, e_j)$ at the point $x$. Hence, using the fact that $X$ is normal to $\mathcal{S}$, we obtain, still working at the point $x$:

$$\frac{d}{dt} g(n\vec{H}, X) = \frac{d}{dt} \sum_{i,j=1}^{n} g^{ij} g(h(e_i, e_j), X)$$

$$= \sum_{i,j=1}^{n} \frac{dg^{ij}}{dt} g(h(e_i, e_j), X) + g^{ij} g\left(\frac{d}{dt} D_{e_i} e_j, X\right)$$

$$+ g^{ij} g\left(h(e_i, e_j), \frac{dX}{dt}\right).$$

Observe that $\sum_{i,j=1}^{n} g^{ij}\left(h(e_i, e_j), \frac{dX}{dt}\right) = g\left(n\vec{H}, \frac{dX}{dt}\right)$, so the last term in the expression above vanishes at $t = 0$. Therefore, using Lemma 4 and the fact that $[g^{ij}]_{1 \le i, j \le n} = diag(\epsilon_1, ..., \epsilon_n)$ at $t = 0$, we get

$$\frac{d}{dt} g(n\vec{H}, X)\bigg|_{t=0} = 2 \sum_{i,j=1}^{n} \epsilon_i \epsilon_j \left(g(h(e_i, e_j), X)\right)^2 + \sum_{i=1}^{n} \epsilon_i g(D_X D_{e_i} e_i, X).$$

We claim that the first term in the expression above is $2g(A_X, A_X)$. To see this, we use the formula $g(X, Y) = \sum_{i=1}^{n} \epsilon_i g(X, e_i) g(Y, e_i)$ to get

$$2 \sum_{i,j=1}^{n} \epsilon_i \epsilon_j \left(g(h(e_i, e_j), X)\right)^2 = 2 \sum_{i,j=1}^{n} \epsilon_i \epsilon_j \left(g(A_X e_i, e_j)\right)^2$$

$$= 2 \sum_{i=1}^{n} \epsilon_i g(A_X e_i, A_X e_i)$$

$$= 2g(A_X, A_X).$$

To compute the second term, we use the definition of the curvature of a connection and again the fact that $[e_i, X] = 0$, so that

$$
\begin{aligned}
D_X D_{e_i} e_i &= -R(X, e_i)e_i + D_{e_i} D_X e_i \\
&= -R(X, e_i)e_i + D_{e_i} D_{e_i} X \\
&= -R(X, e_i)e_i + D_{e_i}\left(-A_X e_i + \nabla^\perp_{e_i} X\right) \\
&= -R(X, e_i)e_i - \nabla_{e_i}(A_X e_i) + h(A_X e_i, e_i) \\
&\quad - A_{\nabla^\perp_{e_i} X} e_i + \nabla^\perp_{e_i} \nabla^\perp_{e_i} X.
\end{aligned}
$$

Hence, using again the fact that $X$ is normal,

$$
\begin{aligned}
g(D_X D_{e_i} e_i, X) &= -g(R(X, e_i)e_i, X) - g(h(A_X e_i, e_i), X) + g(\nabla^\perp_{e_i} \nabla^\perp_{e_i} X, X) \\
&= -g(R(X, e_i)e_i, X) - g(A_X e_i, A_X e_i) + g(\nabla^\perp_{e_i} \nabla^\perp_{e_i} X, X).
\end{aligned}
$$

It follows that

$$
\sum_{i=1}^n \epsilon_i \left(g(D_X(D_{e_i} e_i), X)\right) = -g(R^\perp X, X) - g(A_X, A_X)
$$
$$
+ \sum_{i=1}^n \epsilon_i g(\nabla^\perp_{e_i} \nabla^\perp_{e_i} X, X).
$$

In order to give an intrinsic expression of the last term, we introduce the real function $u := \frac{1}{2} g(X, X)$ and compute, using again the fact that $\nabla_{e_i} e_j(x) = 0$,

$$
\Delta u(x) = \sum_{i=1}^n \epsilon_i \left(g(\nabla^\perp_{e_i} \nabla^\perp_{e_i} X, X) + g(\nabla^\perp_{e_i} X, \nabla^\perp_{e_i} X)\right),
$$

hence

$$
\frac{d}{dt} g(n\vec{H}, X)\bigg|_{t=0} = -g(R^\perp X, X) + g(A_X, A_X) + \Delta u - g(\nabla^\perp X, \nabla^\perp X).
$$

This expression is intrinsic, i.e. there is no more reference to the special orthonormal frame $(e_1, ..., e_n)$, so it holds at any point $x$ of $U$. To conclude the proof, it remains to use the corollary of the divergence theorem (Corollary 1) with the function $u$, taking into account that the variation is local, so that

$$
\frac{d^2}{dt^2} \mathrm{Vol}(\mathcal{S}_t)\bigg|_{t=0} = \int_{\mathcal{S}} \left(g(\nabla^\perp X, \nabla^\perp X) + g(R^\perp X, X) - g(A_X, A_X)\right) dV. \qquad \square
$$

## 1.4   Exercises

(1) Let $S$ be a totally geodesic submanifold of a pseudo-Riemannian manifold $(\mathcal{M}, g)$ and $\gamma$ a regular curve contained $S$. Prove that $\gamma$ is a geodesic of $(\mathcal{M}, g)$ if and only if it is a geodesic of $S$ for the induced metric.

(2) Let $S$ be a non-degenerate submanifold and fix a normal vector field $\xi$. We recall that the shape operator $A_\xi$ is a linear endomorphism of $T_x S$. We define $P_\xi$ to be its characteristic polynomial: $P_\xi(\lambda) = \det(A_\xi - \lambda Id)$. In particular, the roots of $P_\xi$ are the principal curvatures of $A_\xi$;

  (i) Prove that the coefficient of degree 1 of $P_\xi$ is $g(\vec{H}, \xi)$;

  (ii) A non-degenerate submanifold is said to be *austere* if, for all $\xi \in N_x S$, all the coefficients of odd degree of $P_\xi$ vanish. Prove that a surface is austere if and only if it is minimal, that an austere submanifold is minimal but that the converse is not true (hint: many counter-examples may be found in Chapter 3);

  (iii) Assume that $P_\xi$ is diagonalisable. Prove that $S$ is austere if and only if the set of principal curvatures is invariant by multiplication by $-1$, i.e. it is of the form

$$(a, -a, b, -b, ..., c, -c, 0, ..., 0)$$

  (see [Harvey, Lawson (1982)]);

  (iv) A non-degenerate hypersurface with unit normal vector $N$ is said to be *isoparametric* if the coefficients of $P_N$ do not depend on the point $x$. Prove that if $S$ is an isoparametric hypersurface and $P_N$ is diagonalisable, then all its principal curvatures are constant.

# Chapter 2

# Minimal surfaces in pseudo-Euclidean space

In this chapter, we address the study of submanifolds of dimension two, i.e. surfaces, in pseudo-Euclidean space. Beyond the case of geodesics, this is the simplest situation in which the general theory developed in Chapter 1 applies. We first see how the formula for the mean curvature vector simplifies under the two assumptions that the dimension of the submanifold is two and the ambient space is flat. We also show that a pseudo-Riemannian surface admits special coordinates, called *isothermic,* in which the metric takes a simple form. In Section 2.2 we write the partial differential equation that must be satisfied by a function whose graph is a minimal surface in three-dimensional space, and we discuss some examples. In Section 2.3 we give a characterization of those minimal surfaces in pseudo-Euclidean space which are made up of straight lines (ruled surfaces). The last section gives a local description of all minimal surfaces of pseudo-Euclidean space.

## 2.1 Intrinsic geometry of surfaces

Let $(s, t)$ be local coordinates on a surface $\Sigma$ and $g$ a pseudo-Riemannian metric on $\Sigma$. The causal character of the metric is determined by the sign of the determinant $\det g = g_{11}g_{22} - g_{12}^2$ of the metric:

(i) If $\det g = 0$, the metric is degenerate on $T\Sigma$; as we have said in Chapter 1, we will not consider this situation; such surfaces are sometimes called *isotropic;*

(ii) If $\det > 0$, the induced metric is definite. If $S$ is an immersed surface of Minkowski three-space $(\mathbb{R}^3, \langle ., . \rangle_1)$, the metric is necessarily positive and $S$ is usually said to be *spacelike;*

(iii) If $\det g < 0$, the induced metric has signature $(1,1)$ and is in particular indefinite. Surfaces of Minkowski space with indefinite metric are often called *Lorentzian* or *timelike*, however the second terminology is confusing: such a surface admits both timelike and spacelike tangent vectors.

Next the coefficients of the inverse matrix of the first fundamental form are:

$$g^{11} = \frac{g_{22}}{\det g}, \qquad g^{12} = -\frac{g_{12}}{\det g} \quad \text{and} \quad g^{22} = \frac{g_{11}}{\det g}.$$

It follows that the mean curvature vector has the following expression:

$$\vec{H} = \frac{1}{2}\left(\frac{h_{11}g_{22} + h_{22}g_{11} - 2h_{12}g_{12}}{\det g}\right). \tag{2.1}$$

If the first fundamental form takes a simple form, for example if $g_{12}$ vanishes, the mean curvature vector becomes simpler to calculate. The next theorem shows that it is always possible to consider local coordinates such that not only $g_{12} = 0$, but also $|g_{11}| = |g_{22}|$. This result will be useful in Section 2.4 and in Chapter 5. We must however point out that this result is abstract in nature and that in many situations (such as those of Sections 2.2 and 2.3 for example), to write explicit isothermic coordinates is not easy or of no help. The existence of isothermic coordinates should be seen as the two-dimensional analog of the existence of an arc length parameter for curves; there is in general no such kind of "nice" coordinates in higher dimension.

**Theorem 6.** *Let* $(\Sigma, g)$ *be a pseudo-Riemannian surface (i.e. a pseudo-Riemannian manifold of dimension two). Then in the neighbourhood of any point there exists coordinates* $(s, t)$ *which are* isothermic, *i.e.*

$$g_{11} = \epsilon g_{22} \quad and \quad g_{12} = 0,$$

*where* $\epsilon = 1$ *if the metric is definite and* $\epsilon = -1$ *if it is indefinite.*

*Proof.* We only give a sketch of the proof, showing how this geometric problem amounts to solving a Partial Differential Equation (PDE), whose nature depends on the signature of the metric. It is interesting to observe that the proof is somewhat easier in the indefinite case, where the underlying equation is hyperbolic, than in the definite case, where an elliptic PDE must me solved. This dichotomy between the definite and indefinite cases will appear again in the study of minimal surfaces (see Section 2.4.3 at the end of this chapter and Remark 24 of Chapter 5).

*Definite case*:

Let $r$ be a smooth function from an open subset $U$ of $\Sigma$ and introduce the metric $g_0 := e^{2r}g$. A computation shows that the Gaussian curvature $K^{g_0}$ (see Section 1.1.2.3) of $\Sigma$ with respect to $g_0$ is related to the curvature $K^g$ with respect to $g$ by the formula

$$K^{g_0} = e^{2r}(K^g - \Delta_g r).$$

It is known that the elliptic equation $\Delta_g r = K^g$ admits local solutions. It follows that we can choose the function $r$ such that the metric $g_0$ is flat. There is no loss of generality in assuming that $g$ is positive, hence $g_0$ is positive as well. It is then a classical result (see [do Carmo (1992)]) that a flat, positive surface is locally isometric to a piece of the Euclidean plane, so that there exist local coordinates $(s, t)$ such that $g_0 = ds^2 + dt^2$. Hence $g = e^{-2r}(ds^2 + dt^2)$, i.e. $(s, t)$ are local isothermic coordinates for the metric $g$.

*Indefinite case*:

At any point $x$ of $\Sigma$ there exists two independent null vectors, and since the metric is smooth, it implies the existence of two null vector fields $X$ and $Y$ in a neighbourhood of $x$. We claim that there exists vector fields $\tilde{X}$ and $\tilde{Y}$ collinear to $X$ and $Y$ respectively, such that $[\tilde{X}, \tilde{Y}] = 0$: to see this, we set $\tilde{X} = fX$ and $\tilde{Y} = gY$ and compute

$$\begin{aligned}
[fX, gY] &= \nabla_{fX} gY - \nabla_{gY} fX \\
&= f(g\nabla_X Y + X(g)Y) + g(f\nabla_Y X + Y(f)X) \\
&= fg[X, Y] + X(g)Y - Y(f)X.
\end{aligned}$$

Writing the vanishing of $[fX, gY]$ in the basis $(X, Y)$ gives

$$\begin{cases} \lambda fg + Y(f) = 0 \\ \mu fg + X(g) = 0, \end{cases}$$

where $\lambda$ and $\mu$ are the coefficient of $[X, Y]$ in the basis $(X, Y)$, i.e. $[X, Y] = \lambda X + \mu Y$. This is a first order, hyperbolic system of partial differential equations in terms of the unknown functions $\lambda$ and $\mu$, which admits local solutions (see [Weinstein (1996)]). The condition $[\tilde{X}, \tilde{Y}] = 0$ means that the flows of $\tilde{X}$ and $\tilde{Y}$ commute, i.e. there is a local system of coordinates $(u, v)$ such that $\partial_u = \tilde{X}$ and $\partial_v = \tilde{Y}$. The coordinates $(u, v)$ are called *null* coordinates for obvious reasons. Once we have null coordinates, it is easy to get isothermic coordinates: we set $Z := \tilde{X} + \tilde{Y}$ and $W := \tilde{X} - \tilde{Y}$, so that

$[Z, W] = 2[\tilde{Y}, \tilde{X}] = 0$ and we still have commuting flows. Calling $(s, t)$ the resulting system of coordinates, we see that

$$g(\partial_s, \partial_s) = g(Z, Z) = 2g(\tilde{X}, \tilde{Y}),$$
$$g(\partial_t, \partial_t) = g(W, W) = -2g(\tilde{X}, \tilde{Y}),$$

and

$$g(\partial_s, \partial_t) = g(Z, W) = 0,$$

which completes the proof. $\qquad\qquad\qquad\qquad\qquad\qquad\qquad\qquad$ $\square$


## 2.2   Graphs in Minkowski space

In this section we consider the space $\mathbb{R}^3$ with coordinates $(x_1, x_2, x_3)$ and metric

$$\langle \cdot, \cdot \rangle_1 = dx_1^2 + dx_2^2 - dx_3^2.$$

Here we deviate from the convention adopted in the rest of the book, where the negative directions are put before the positive ones in the labeling of the coordinates: it is visually more comfortable that the distinguished direction, so here the negative one, be vertical. The simplest way to describe a submanifold is as a graph of a function. Moreover, this can be done locally without lost generality: at a neighbourhood of a point $x$, a submanifold $\mathcal{S}$ can be described a graph on his tangent space $T\mathcal{S}$. We thus consider

$$\mathcal{S} := \{(x_1, x_2, u(x_1, x_2)) \in \mathbb{R}^3 \,|\, (x_1, x_2) \in U\},$$

where $U$ is an open subset of $\mathbb{R}^2$ and $u$ is smooth, real function on $U$. In other words, $\mathcal{S}$ is the image of the immersion $f(x_1, x_2) = (x_1, x_2, u(x_1, x_2))$. The first derivatives of $f$ provide a basis $(X_1, X_2)$ of tangent vectors to $\mathcal{S}$:

$$X_1 := f_{x_1} = (1, 0, u_{x_1}), \qquad X_2 := f_{x_2} = (0, 1, u_{x_2}).$$

Here and in the remainder of the book, we shall denote a partial derivative by the corresponding variable put in index. The first fundamental form of $\mathcal{S}$ is given by

$$g_{11} = 1 - u_{x_1}^2, \qquad g_{12} = -u_{x_1} u_{x_2} \quad \text{and} \quad g_{22} = 1 - u_{x_2}^2$$

so we have

$$\det g = 1 - u_{x_1}^2 - u_{x_2}^2 = 1 - |\nabla u|_0^2.$$

In particular $\mathcal{S}$ is spacelike (resp. Lorentzian, degenerate) if $|\nabla u|_0^2 < 1$ (resp. if $|\nabla u|_0^2 > 1$, $|\nabla u|_0^2 = 1$).

We now assume that $|\nabla u|_0^2 \neq 1$ and write down the condition of minimality of $\mathcal{S}$ in terms of the function $u$. For this purpose we need to compute the covariant derivatives of the tangent vectors to $\mathcal{S}$:

$$D_{X_1} X_1 = f_{x_1 x_1} = (0, 0, u_{x_1 x_1}), \qquad D_{X_1} X_2 = f_{x_1 x_2} = (0, 0, u_{x_1 x_2}),$$

$$D_{X_2} X_2 = f_{x_2 x_2} = (0, 0, u_{x_2 x_2}).$$

A surface in $\mathbb{R}^3$ is a hypersurface, hence, as we discussed in Section 1.2.4 of Chapter 1, the second fundamental form is better seen as a scalar rather that a vectorial quantity. A unit normal vector to $\mathcal{S}$ is given by

$$N = \left| 1 - |\nabla u|^2 \right|^{-1/2} (u_{x_1}, u_{x_2}, 1).$$

Thus, setting $W := \left| 1 - |\nabla u|_0^2 \right|^{1/2}$, the coefficients of the second fundamental form are

$$h_{11} = \frac{1}{W} u_{x_1 x_1}, \qquad h_{12} = \frac{1}{W} u_{x_1 x_2} \quad \text{and} \quad h_{22} = \frac{1}{W} u_{x_2 x_2}.$$

Hence from Equation (2.1) we get

$$H = \frac{1}{2} W^{-3} \left( u_{x_1 x_1}(1 - u_{x_2}^2) + u_{x_2 x_2}(1 - u_{x_1}^2) + 2 u_{x_1 x_2} u_{x_1} u_{x_2} \right).$$

In particular, the graph of the function $u$ is minimal if and only if $u$ satisfies:

$$u_{x_1 x_1}(1 - u_{x_2}^2) + u_{x_2 x_2}(1 - u_{x_1}^2) + 2 u_{x_1 x_2} u_{x_1} u_{x_2} = 0. \qquad (2.2)$$

This is a PDE which is non-linear but depends linearly on the second derivatives. Such PDE is called *quasi-linear*. We will not discuss the general theory of existence of solutions to this equation, but rather describe a couple of special solutions. Of course, a function $u$ with vanishing second derivatives (an *affine* function) is solution of Equation (2.2). Although these are not very exciting solutions, the resulting surfaces $\mathcal{S}$ being planes, they are important because of the following uniqueness theorem, proved in [Calabi (1968)]:

**Theorem 7 (Calabi).** *Let $u$ be a solution of Equation (2.2), on $U = \mathbb{R}^2$ (we say that the graph is entire) and such that $|\nabla u|_0 < 1$ (hence $\mathcal{S}$ is spacelike); then $u$ is an affine function.*

There is an analogous theorem (which actually predates Calabi's) in the case of the Riemannian metric $\langle ., .\rangle_0 = dx_1^2 + dx_2^2 + dx_3^2$, and is known as the Bernstein theorem. Both results admit generalizations to hypersurfaces of higher dimensions, but surprisingly, not to any dimension in the Riemannian case: there does exist minimal hypersurfaces in $(\mathbb{R}^m, \langle ., .\rangle_0)$ which are non affine graphs when $m \geq 9$ (see [Bombieri, De Giorgi, Giusti (1969)]). The generalization of Calabi's theorem is due to Calabi itself for $m = 4$ and to [Cheng, Yau (1976)] for higher dimension. We refer to [Nelli, Soret (2005)] and the references therein for an account about the fascinating story of the Bernstein problem in the Riemannian case.

The easiest way of finding at least one or more solutions to a complicated PDE is often to look for solutions which have some symmetry. This reduces the PDE to an ordinary differential equation (abbreviated to ODE) that we hopefully are able to solve. For example, we shall seek a solution $u$ of (2.2) which is radial, i.e. which depends only on $r := \sqrt{x_1^2 + x_2^2}$. In other words, the function $u$ is constant on the circular arcs of $U$.

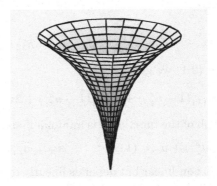

Fig. 2.1   The spacelike elliptic catenoid

Writing $u(x_1, x_2) = v \circ r(x_1, x_2)$ and taking into account that

$$r_{x_1} = \frac{x_1}{r}, \qquad r_{x_2} = \frac{x_2}{r},$$

we have

$$u_{x_1} = \frac{x_1}{r} v', \qquad u_{x_2} = \frac{x_2}{r} v' \tag{2.3}$$

and

$$u_{x_1 x_1} = \frac{x_1^2}{r^2} v'' + \frac{x_2^2}{r^3} v',$$

as well as analogous expressions for the other second derivatives. Then it is not difficult to see that Equation (2.2) reduces to

$$v'' + \frac{v'(1 - (v')^2)}{r} = 0. \tag{2.4}$$

The general solution of (2.4) takes two different forms, depending whether $|v'|$ is less or greater than 1 (we discard the solutions $|v'| \equiv 1$, whose corresponding surfaces, the light cones $\{x_3 = \pm\sqrt{x_1^2 + x_2^2}\}$, are degenerate). Since $|\nabla u| = |v'|$, the first type of solutions will provide a spacelike surface, while the second type will give an indefinite surface. The first family of solutions is $v'(r) = (1 + (r/c)^2)^{-1/2}$, where $c$ is some non-vanishing, real constant. It remains to integrate to find $v(r) = c\sinh^{-1}(r/c) + c_0$, where $c_0$ is another constant. We argue that besides the existence of two constants of integration, we have obtained essentially one surface: varying $c$ amounts to a homothety of the surface centered at the origin of $\mathbb{R}^3$, while varying the constant $c_0$ consists of translating the surface along the vertical coordinate axis $x_3$. We observe that the function $u$ fails to be differentiable at the origin $r = 0$, since $v'(0) = 1$ and taking into account (2.3). More precisely, the graph of $u$ is asymptotic to the light cone $\{x_3 = \sqrt{x_1^2 + x_2^2}\}$ at the origin. In particular, this example shows that the assumption "entire" in Calabi's theorem is necessary: we have just described a spacelike, minimal graph on $\mathbb{R}^2 - \{0\}$, which is however not affine. Because of the special form of $u = v(\sqrt{x_1^2 + x_2^2})$, the corresponding graph $S$ is foliated by horizontal circles centered on the vertical coordinate axis, hence it is a surface of revolution. In analogy with the similar surface of the Riemannian three-space (cf Exercise 1), we call it *spacelike elliptic catenoid*[1] (the use of the word elliptic will become clearer in a moment).

The second family of solutions of (2.4) is $v'(r) = (1 - (r/c)^2)^{-1/2}$, where $c$ is some non-vanishing constant. Integrating, it gives $v(r) = c\sin^{-1}(r/c) + c_0$, where $c_0$ is another constant. Again, the constants $c$ and $c_0$ correspond respectively to homotheties and vertical translations of the surface. The asymptotic behaviour of the corresponding function $u_{c,c_0}$ at the origin is similar to the one of the first case, however, this solution exists only for $r \in (0, c)$. At $r = c$, the graph of $u_{c,c_0}$ becomes vertical and touches the horizontal plane of equation $x_3 = c\pi/2 + c_0$. The reflection of $u_{c,c_0}$ about this plane is the graph of $-u_{c,c\pi+c_0}$. The surface obtained taking the union of the graphs of $u_{c,c_0}$ and $-u_{c,c\pi+c_0}$ is topologically a sphere punctured with two singularities. We shall call it *Lorentzian elliptic catenoid*.

In the next chapter, not only we shall see how to obtain these two surfaces in a more geometric way, but also generalize the construction to higher dimension, and even to more general ambient spaces than the Euclidean one.

---

[1] This surface is called *catenoid of the first kind* in [Kobayashi (1983)].

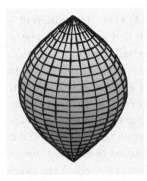

Fig. 2.2   The Lorentzian elliptic catenoid

Analogously, we may seek a solution $u$ which depends only on the argument of $x_1 + ix_2$. Let $U$ be a simply connected open subset of $\mathbb{R}^2 - \{0\}$. Then we may assign a unique value to the argument of any complex number $x_1 + ix_2$, where $(x_1, x_2) \in U$. Denoting by $arg(x_1 + ix_2)$ this argument function, we set $u(x_1, x_2) := v \circ \arg(x_1 + ix_2)$ on $U$. Differentiating the identity

$$x_1 + ix_2 = re^{i \arg(x_1 + ix_2)},$$

we find that

$$\arg(x_1 + ix_2)_{x_1} = \frac{-x_2}{r^2} \quad \text{and} \quad \arg(x_1 + ix_2)_{x_2} = \frac{x_1}{r^2}.$$

Hence

$$u_{x_1} = \frac{-x_2}{r^2} v' \quad \text{and} \quad u_{x_2} = \frac{x_1}{r^2} v'. \tag{2.5}$$

Next we get

$$u_{x_1 x_1} = \frac{x_2^2}{r^4} v'' + \frac{2x_1 x_2}{r^4} v'$$

and analogous expressions for the other second derivatives. Finally, Equation (2.2) reduces to $v'' = 0$, so the solution we are looking for is

$$u(x_1, x_2) = c \arg(x_1 + ix_2) + c_0,$$

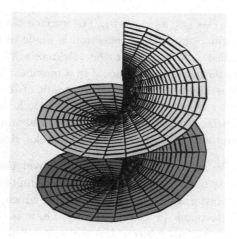

Fig. 2.3   The elliptic helicoid

where $c$ and $c_0$ are two real constants. We may assume that $c$ does not vanish, since otherwise $u$ is constant and the resulting surface is a horizontal plane.

We see in particular that

$$|\nabla u|^2 = \frac{c^2}{x_1^2 + x_2^2}.$$

It follows that the graph $\mathcal{S}$ of $u$ has several causal characters: it is degenerate on the intersection with the vertical cylinder $C := \{x \in \mathbb{R}^3, x_1^2 + x_2^2 = c^2\}$, while it is spacelike (resp. indefinite) outside $C$ (resp. inside $C$). We furthermore observe that $\mathcal{S}$ is the union of horizontal straight half lines: the intersection of $\mathcal{S}$ with a horizontal plane of height $x_3 = u$ is simply the half-ray of equation $\arg(x_1 + ix_2) = (u - c_0)/c$.

Of course, $u$ cannot be extended to the whole $\mathbb{R}^2 - \{0\}$ as a single-valued function, but we may do so as a multi-valued one: we consider all the points $(x_1, x_2, x_3)$ of $\mathbb{R}^3$ such that $x_3$ is any possible argument of $(u - c_0)/c$. Hence we get a surface which is invariant by the vertical translation $(x_1, x_2, x_3) \mapsto (x_1, x_2, x_3 + 2\pi)$. However we are not quite satisfied with the surface we have obtained in this way, since it is not complete: its boundary is the vertical axis $\{x_1 = 0, x_2 = 0\}$. This suggests to consider the reflection of the surface about this axis, which is actually the same surface translated by a vertical translation $(x_1, x_2, x_3) \mapsto (x_1, x_2, x_3 + \pi)$. The surface obtained by taking the union of the original one with it reflection is made of horizontal lines and is classically called *helicoid*. We rather nickname it *elliptic helicoid*, since we shall meet another kind of helicoid in a moment. The importance of this last example is highlighted by a recent result of [Fernández, Lopez (2010)] claiming that the spacelike part of the elliptic helicoid is the only spacelike minimal surface satisfying certain natural assumptions. This is thus an indefinite counterpart of the theorem of [Meeks, Rosenberg (2005)].

Although spacelike surfaces are always graphs over the horizontal coordinate plane, this is not necessarily so for an indefinite surface. Hence we now turn our attention to the case of a surface which is a graph over the vertical coordinate plane $\{x \in \mathbb{R}^3, x_1 = 0\}$, i.e. it is parametrized by $f(x_2, x_3) = (u(x_2, x_3), x_2, x_3)$, where $u$ is again a smooth, real function on some open subset of $\mathbb{R}^2$. The first derivatives of $f$ are

$$f_{x_2} = (u_{x_2}, 1, 0) \quad \text{and} \quad f_{x_3} = (u_{x_3}, 0, 1).$$

Hence the first fundamental form is given by

$$g_{11} = u_{x_2}^2 + 1, \qquad g_{12} = u_{x_2} u_{x_3}, \quad \text{and} \quad g_{22} = u_{x_3}^2 - 1,$$

so we have

$$\det g = u_{x_3}^2 - u_{x_2}^2 - 1 = -|\nabla u|_1^2 - 1.$$

Here $|.|_1^2$ denotes the squared norm of the indefinite metric $\langle ., . \rangle_1 = dx_2^2 - dx_3^2$, which is the induced metric on the plane $\{x \in \mathbb{R}^3, x_1 = 0\}$. We deduce that $\mathcal{S}$ is spacelike (resp. Lorentzian, degenerate) if $|\nabla u|_1^2 < -1$ (resp. if $|\nabla u|_1^2 > -1$, $|\nabla u|_1^2 = -1$).

Assume now that $|\nabla u|_1^2 \neq -1$. The second derivatives of $f$ are

$$f_{x_2 x_2} = (u_{x_2 x_2}, 0, 0), \qquad f_{x_2 x_3} = (u_{x_2 x_3}, 0, 0) \quad \text{and} \quad f_{x_3 x_3} = (u_{x_3 x_3}, 0, 0),$$

and a unit normal vector to $\mathcal{S}$ is given by

$$N = \left|1 + |\nabla u|_1^2\right|^{-1/2} (1, u_{x_2}, u_{x_3})$$

so, setting $W := \left|1 + |\nabla u|_1^2\right|^{1/2}$, we have, by Equation (2.1):

$$H = \frac{1}{2} W^{-3} \left( u_{x_2 x_2}(u_{x_3}^2 - 1) + u_{x_3 x_3}(u_{x_2}^2 + 1) - 2 u_{x_2 x_3} u_{x_2} u_{x_3} \right)$$

and the graph of the function $u$ is minimal if and only if $u$ satisfies:

$$u_{x_2 x_2}(u_{x_3}^2 - 1) + u_{x_3 x_3}(u_{x_2}^2 + 1) - 2u_{x_2 x_3}u_{x_2}u_{x_3} = 0. \tag{2.6}$$

Again, we see that affine functions provide easy but uninteresting solutions of Equation (2.6) (*trivial* solutions), but we will find more interesting ones. Set $U := \{(x_2, x_3) \in \mathbb{R}^2, x_2^2 - x_3^2 \neq 0\}$ and look for a solution of the form $u(x_2, x_3) = v \circ r(x_2, x_3)$, where $r = \epsilon\sqrt{|x_2^2 - x_3^2|}$. Observe that the level sets of the function $r$ are the hyperbolas which should be regarded as the Lorentzian analogs of the circles $r = \sqrt{x_1^2 + x_2^2}$ considered before, since they are curves of constant curvature for the metric $dx_2^2 - dx_3^2$.

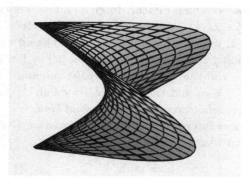

Fig. 2.4 The spacelike hyperbolic catenoid

We now compute the first derivatives of $u = v \circ r$: taking into account that

$$r_{x_2} = \epsilon\frac{x_2}{r} \qquad r_{x_3} = -\epsilon\frac{x_3}{r},$$

where we have set $\epsilon := \frac{x_2^2 - x_3^2}{|x_2^2 - x_3^2|}$, we have

$$u_{x_2} = \epsilon\frac{x_2}{r}v', \qquad u_{x_3} = -\epsilon\frac{x_3}{r}v'. \tag{2.7}$$

In particular,

$$\det g = u_{x_3}^2 - u_{x_2}^2 - 1 = \frac{x_2^2 - x_3^2}{r^2}(v')^2 - 1 = \epsilon(v')^2 - 1,$$

which shows that if $\epsilon = -1$, the surface is indefinite, while if $\epsilon = 1$, the causal character depends on whether $|v'|$ is less or greater than 1. Next we compute the second derivatives of $u$:

$$u_{x_2 x_2} = \frac{x_2^2}{r^2} v'' - \frac{x_3^2}{r^3} v',$$

as well as analogous expressions for the other second derivatives. Then Equation (2.6) reduces to

$$\epsilon v'' + \frac{v'((v')^2 + \epsilon)}{r} = 0. \tag{2.8}$$

If $\epsilon = 1$, the general solution of (2.8) is $v'(r) = ((r/c)^2 - 1)^{-1/2}$, where $c$ is some non-vanishing constant. Integrating, we get to $v(r) = c \cosh^{-1}(r/c) + c_0$, where $c_0$ is another constant. With the multiplication of interesting examples, it is now becoming difficult to give names! We call the graph of $u = c \cosh^{-1}(\sqrt{x_2^2 - x_3^2}/c) + c_0$ the *Lorentzian hyperbolic catenoid of the first kind*. If $\epsilon = -1$, we recover Equation (2.4), so again we have two sorts of solutions (excluding the degenerate solution $|v'| = 1$ which again gives the light-cone). We call the *spacelike hyperbolic catenoid*[2] the graph $x_1 = c \sin^{-1}(\sqrt{x_3^2 - x_2^2}/c) + c_0$ and the graph $x_1 = c \sinh^{-1}(\sqrt{x_3^2 - x_2^2}/c) + c_0$ the *Lorentzian hyperbolic catenoid of the second kind*.

There is one more interesting solution to Equation (2.6), that we give without saying from where it comes:

$$u := \tanh^{-1} \frac{x_3}{x_2}.$$

The reader will check easily that it is solution of Equation (2.6) and that it is also made up of half-rays, as is the elliptic helicoid. We call the graph of $u$ the *hyperbolic helicoid*[3] and we refer the reader to the next section for a geometric discussion about these two surfaces.

## 2.3　The classification of ruled, minimal surfaces

A *ruled surface* of $\mathbb{R}^m$ is a surface made up of points belonging to a one-parameter family of straight lines. The purpose of this section is to give a characterization of minimal, ruled surfaces in $(\mathbb{R}^m, \langle ., . \rangle_p)$.

A straight line may be parametrized by $t \mapsto \gamma t + x$, where $\gamma$ is a non-vanishing vector and $x$ is a point. We have some freedom in the choice of the direction $\gamma$ and the base point $x$: we may replace $\gamma$ by any vector collinear to it. In particular, if $\gamma$ is not null, we may assume that it is a

---

[2] This surface is called *catenoid of the second kind* in [Kobayashi (1983)].
[3] This surface is called *helicoid of the second kind* in [Kobayashi (1983)].

unit vector, i.e. $|\gamma|_p^2 = 1$ or $-1$. Moreover, we may translate the point $x$ along the line, i.e. replace it by $\tilde{x} := x + \lambda\gamma$ where $\lambda$ is a real parameter.

We first discuss the easier case where all of the rulings are parallel. In this case the surface is sometimes called a *cylinder* and admits a parametrization of the form

$$f : I \times \mathbb{R} \to \quad \mathbb{R}^m$$
$$(s, t) \mapsto \gamma_0 t + x(s),$$

where $x$ is a curve in $\mathbb{R}^m$, and $\gamma_0$ a fixed direction. A short calculation shows that $f_{st}$ and $f_{tt}$ vanish, so $h_{12}$ and $h_{22}$ vanish as well; we deduce that $\vec{H} = \frac{h_{11}g_{22}}{\det g}$. It follows that there are two cases in which the immersion is minimal: if $h_{11}$ vanishes, the whole tensor $h$ vanishes and the surface is totally geodesic, hence a plane (see Proposition 4 in Chapter 1). The situation is more interesting if $g_{22} = |\gamma_0|_p^2$ vanishes. It means that the direction of the ruling $\gamma_0$ is null. If this is the case, the non-degeneracy assumption reads $\det g = g_{12}^2 = (\langle x', \gamma_0 \rangle_p)^2 \neq 0$. We claim that if this condition holds, we may reparametrize the curve $x$ to make it a null curve, i.e. in such a way that $|x'|_p^2 = 0$. To see this, set

$$\tilde{x} = x + \lambda\gamma_0,$$

so that

$$|\tilde{x}'|_p^2 = |x'|_p^2 + 2\lambda'\langle x', \gamma_0 \rangle_p.$$

Hence, choosing $\lambda = -\int_s \frac{|x'|_p^2}{2\langle x', \gamma_0 \rangle_p}$ makes $\tilde{x}$ a null curve and in particular $(s, t)$ are null coordinates for the immersion $\tilde{f} = \gamma_0 t + \tilde{x}$. Hence we have a family of minimal, ruled surfaces: take a null direction $\gamma_0$ and a null curve $x$ in $(\mathbb{R}^m, \langle ., . \rangle_p)$ such that $\langle x', \gamma_0 \rangle_p \neq 0$. Then the immersion $f(s, t) = \gamma_0 t + x(s)$ is minimal.

We now assume that the rulings are not parallel, so that the ruled surface may be locally parametrized by an immersion of the form

$$f : I \times \mathbb{R} \to \quad \mathbb{R}^m$$
$$(s, t) \mapsto \gamma(s) t + x(s),$$

where $\gamma$ and $x$ are two curves in $\mathbb{R}^m$, and $\gamma$ is moreover assumed to be regular. We first observe that in this setting, that the direction of the ruling cannot be null:

**Lemma 5.** *Let $S$ be a ruled, minimal surface. If the direction of the rulings is not constant, then it is never null.*

**Remark 6.** At this point we should warn the reader about a possible confusion: what we claim is that the direction, i.e. the *position vector* of the curve $\gamma$ is not null: $|\gamma|_p^2 \neq 0$. It may very well happen (and we will see that it does so) that the curve $\gamma$ is a null curve, i.e. its *velocity vector* is null: $|\gamma'|_p^2 = 0$.

*Proof.* We proceed by contradiction, assuming that we have indeed $|\gamma|_p^2 = 0$. Rather than computing all of the coefficients of the first and second fundamental forms, observe that $f_{tt}$ vanishes, hence both $g_{22} = |\gamma|_p^2$ and $h_{22}$ vanish. Thus Equation (2.1) implies that the immersion is minimal if and only if $h_{12}$ vanishes, i.e. if $f_{st} = \gamma'$ belongs to the span of $f_s = \gamma't + x'$ and $f_t = \gamma$, so that there exist functions $\lambda$ and $\mu$ such that

$$\gamma' = \lambda(\gamma't + x') + \mu\gamma$$

By the assumption that the directions of the rulings are not constant, $\lambda \neq 0$ (otherwise, we would have $\gamma = \exp(\int \mu)e_1$ where $e_1$ is a fixed direction). So we get

$$x' = \frac{1-t\lambda}{\lambda}\gamma' - \frac{\mu}{\lambda}\gamma.$$

Taking into account that $\langle \gamma, \gamma' \rangle_p = \frac{1}{2}\frac{d|\gamma|_p^2}{ds} = 0$, we deduce that

$$g_{12} = \langle \gamma, \gamma' + tx' \rangle_p = 0,$$

which implies that the first fundamental form is degenerate, a contradiction. □

We come back to our description of a ruled surface with non-parallel rulings. As in the case of parallel rulings, we may

(i) replace the vector $\gamma$ by a vector collinear to it; therefore we assume from now on that $g_{22} = |\gamma|_p^2 = \epsilon$, where $\epsilon = 1$ or $-1$;

we may also

(ii) translate the point $x(s)$ along the ruling, i.e. replace it by $\tilde{x}(s) := x(s) + \lambda(s)\gamma(s)$ where $\lambda$ is a real function. Moreover, if $|\gamma'|_p^2 \neq 0$, we may assume that $s$ is the arc length parameter of $\gamma$. Since the discussion is local, we have $|\gamma'|_p^2$ is equal to a constant $\eta$ equal to 1, $-1$ or 0. The corresponding surfaces will be referred as to *elliptic, hyperbolic* and *parabolic,* respectively.

We claim that, using point (ii) above, we may reparametrize the surface $\mathcal{S}$ in such a way that the coefficient $g_{12}$ of the first fundamental form vanishes: replacing $x(s)$ by $\tilde{x}(s) := x(s) + \lambda(s)\gamma(s)$, we get

$$g_{12} = \langle f_s, f_t \rangle = \langle \gamma, x' \rangle_p + \epsilon\lambda',$$

hence, choosing $\lambda = -\epsilon \int_s \langle \gamma, x' \rangle_p$, we obtained the claim.

Next, we compute the second derivatives of the immersion $f$:

$$f_{ss} = \gamma'' t + x'' \qquad f_{st} = \gamma' \qquad f_{tt} = 0.$$

In particular, the coefficient $h_{22}$ vanishes as well and Equation (2.1) becomes:

$$2\vec{H} = \frac{h_{11}}{g_{11}}.$$

Thus, the immersion $f$ is minimal if and only if the coefficient $h_{11}$ vanishes, i.e. $f_{ss}$ belongs to the span of $f_s$ and $f_t$. Since $f_s$ and $f_t$ are orthogonal, it follows that the basis $(f_s/|f_s|_p^2, f_t/|f_t|_p^2)$ is orthonormal, and we can write

$$f_{ss} = \frac{\langle f_{ss}, f_s \rangle_p}{|f_s|_p^2} f_s + \frac{\langle f_{ss}, f_t \rangle_p}{|f_t|_p^2} f_t.$$

Recalling that both $|\gamma|_p^2$ and $|\gamma'|_p^2$ are constant, we have

$$\langle \gamma', \gamma'' \rangle_p = 0 \quad \text{and} \quad \langle \gamma, \gamma'' \rangle_p = -|\gamma'|_p^2 = -\eta.$$

By an easy computation we deduce

$$f_{ss} = \frac{(\langle \gamma'', x' \rangle_p + \langle \gamma', x'' \rangle_p) t + \langle x'', x' \rangle_p}{\eta t^2 + 2\langle \gamma', x' \rangle_p t + |x'|_p^2} (\gamma' t + x') + \epsilon(-\eta t + \langle \gamma, x'' \rangle_p) \gamma.$$

The left hand side is $f_{ss} = \gamma'' t + x''$, a polynomial in the variable $t$ of degree one, so, since $\gamma' \neq 0$, the rational fraction in the right hand side of the expression above must simplify to a polynomial of degree zero, i.e. an expression not depending on $t$. We shall denote this by $C(s)$. Equating the constant coefficient and the coefficient of degree one on both sides, we get

$$\begin{cases} \gamma'' = C(s)\gamma' - \epsilon \eta \gamma \\ x'' = C(s)x' + \epsilon \langle \gamma, x'' \rangle_p \gamma. \end{cases}$$

The first equation allows us to determine completely the curve $\gamma$. Before we do this, we first claim that the coefficient $C$ must vanish. If $\eta = 1$ or $\eta = -1$, this is a mere consequence of the Frénet equation (see Chapter 1, Section 1.2.3.3), but we have to check this in the case of vanishing $\eta = 0$. In such a case, the general solution of the equation $\gamma'' = C(s)\gamma'(s)$, is

$$\gamma'(s) = e^{\int_s C(s)} e_1,$$

where $e_1$ is a null vector (since $|\gamma'|_p^2 = |e_1|_p^2 = \eta = 0$). So

$$\gamma(s) = \mu(s)e_1 + e_2,$$

where $e_2$ is a non-null vector and $\mu(s)$ a primitive of $e^{\int_s C(s)}$. Hence the curve $\gamma(s)$ is a straight line. On the other hand, since we are in the null case, we have not yet used the freedom that we enjoy in reparametrizing $\gamma(s)$. We do so now, and set $\gamma(s) := e_1 s + e_2$. But this implies $\gamma'' = 0$, hence $C$ vanishes again.

Since $C$ vanishes, we are left to a simple second order equation with constant coefficients: $\gamma'' + \epsilon\eta\gamma = 0$. Its general solution depends on the value of $\epsilon\eta$:

(i) If $\epsilon\eta = 1$, the curve $\gamma$ is a circle, i.e. it takes the form $\gamma(s) = \cos s\, e_1 + \sin s\, e_2$ where $e_1$ and $e_2$ are two unit vectors which are orthogonal and of the same causal character (they are both positive or both negative);

(ii) If $\epsilon\eta = -1$, the curve $\gamma$ is a hyperbola, i.e. it takes the form $\gamma = \cosh s\, e_1 + \sinh s\, e_2$, where $e_1$ and $e_2$ are two unit, orthogonal vectors, one of two distinct causal characters (of course this case does not occur if the metric is positive, i.e. if $p = 0$);

(iii) If $\eta = 0$, $\gamma'' = 0$, so that $\gamma$ a straight line $\gamma(s) = e_0 s + e_1$, where $e_0$ is a null, non-vanishing vector (if $e_0$ vanishes, we fall back in the case already treated of the ruling having constant direction). The condition $|\gamma|_p^2 = \epsilon$ implies furthermore that $|e_1|_p^2 = \epsilon$ and $\langle e_0, e_1 \rangle_p = 0$.

Once the curve $\gamma$ has been computed explicitly, one can determine $x$ by the second equation of the system above. For this purpose we first need to find out $\langle \gamma, x'' \rangle_p$. Since $C(s) = 0$, we have

$$\langle \gamma'', x' \rangle_p + \langle \gamma', x'' \rangle_p = \frac{d\langle \gamma', x' \rangle_p}{ds} = 0.$$

On the other hand,

$$\frac{dg_{12}}{ds} = \langle \gamma', x' \rangle_p + \langle \gamma, x'' \rangle_p = 0,$$

so we deduce that $\langle \gamma, x'' \rangle_p$ is constant as well.

If $\eta \neq 0$, we have

$$x'' = \epsilon\langle \gamma, x'' \rangle_p \gamma = -\eta\langle \gamma, x'' \rangle_p \gamma''.$$

Integrating, we get $x = -\langle \gamma, x'' \rangle_p \eta\gamma + vs + x_0$, where $v$ and $x_0$ are respectively a constant vector and a fixed point of $\mathbb{R}^m$. As we have seen above we may replace $x(s)$ by $\tilde{x} := vs + x_0$. Moreover, we may assume without loss of generality that $x_0 = 0$. On the other hand, the assumption $0 = g_{12} = \langle x', \gamma \rangle_p = \langle v, \gamma \rangle_p$ implies that $v$ is orthogonal to both $e_1$ and $e_2$. Moreover, the non-degeneracy implies that $v$ must be a non-null vector, hence we write $v = ce_3$, where $c$ is a non-vanishing constant and $e_3$ is a unit vector. Thus the immersion takes the form $f(s, t) = (\cos s\, e_1 + \sin s\, e_2)t + ce_3 s + x_0$ or $f(s, t) = (\cosh s\, e_1 + \sinh s\, e_2)t + ce_3 s + x_0$. Finally, making a suitable translation and scaling, we may set $x_0 = 0$ and $c = 1$.

In the parabolic case $\eta = 0$, we integrate the equation $x'' = \epsilon \langle \gamma, x'' \rangle_p \gamma = c(e_0 s + e_1)$ to get $x = c(\frac{s^3}{3}e_0 + s^2 e_1) + vs + x_0$, where $c$ is real constant, $v$ a constant vector, and $x_0$ a fixed point of $\mathbb{R}^m$. After rescaling and translating the surface if necessary, we may assume that $c = 1$ and $x_0 = 0$. The immersion takes therefore the form

$$f(s, t) = (se_0 + e_1)t + sv + \left( \frac{s^3}{3}e_0 + s^2 e_1 \right)$$

$$= (te_0 + v)s + e_1 t + \left( \frac{s^3}{3}e_0 + s^2 e_1 \right).$$

Observe that replacing $v$ by $\tilde{v} := v + \lambda e_0$, where $\lambda$ is a real constant, yields a different parametrization of the same surface. On the other hand the assumption $\langle x', \gamma \rangle_p = 0$ implies that $\langle e_0, \tilde{v} \rangle_p + 2\epsilon = 0$ and $\langle e_1, v \rangle_p = 0$. Hence for a suitable choice of $\lambda$, we have $e_0 = e_2 + e_3$ and $\tilde{v} = e_2 - e_3$, where $(e_2, e_3)$ is an orthonormal basis of the plane spanned by $v$ and $e_0$, with $|e_2|^2 = \epsilon$.

Summing up, we have proved:

**Theorem 8.** *Let $S$ be a ruled, minimal, non-planar surface of Euclidean space $\mathbb{R}^m$ endowed with the metric $\langle ., . \rangle_p$. Then $S$ is, up to translation and scaling, an open subset of one of the following surfaces:*

(i) *A minimal cylinder*

$$f(s, t) = \gamma_0 t + x(s),$$

*where $\gamma_0$ is a null vector and $x(s)$ is a null curve such that $\langle \gamma_0, x' \rangle_p \neq 0$;*

(ii) *An elliptic helicoid*

$$f(s, t) = (\cos s\, e_1 + \sin s\, e_2)t + e_3 s,$$

*where $(e_1, e_2, e_3)$ is an orthonormal family (i.e. $e_1, e_2$ and $e_3$ are three unit, mutually orthogonal vector), such that $e_1$ and $e_2$ have the same causal character;*

(iii) *A hyperbolic helicoid*

$$f(s, t) = (\cosh s\, e_1 + \sinh s\, e_2)t + e_3 s,$$

*where $(e_1, e_2, e_3)$ is an orthonormal family, such that $e_1$ and $e_2$ have distinct causal characters;*

Fig. 2.5   The parabolic helicoid

(iv) *A parabolic helicoid*[4]

$$f(s,t) = (t + s^2)e_1 + \left(\frac{s^3}{3} + st + s\right)e_2 + \left(\frac{s^3}{3} + st - s\right)e_3$$

*where $(e_1, e_2, e_3)$ is an orthonormal family, such that $e_1$ and $e_2$ have the same causal character and $e_3$ has the opposite one.*

**Remark 7.** The minimal cylinders have always indefinite metric. The causal character of the helicoids is versatile. In the elliptic and hyperbolic cases, we have

$$\det g = (|e_2|_p^2 t^2 + |e_3|_p^2)|e_1|_p^2,$$

so the elliptic helicoid is of course positive if all vectors $e_1, e_2$ and $e_3$ are positive. If the causal character of $e_3$ is distinct from the one of $e_2$, the elliptic helicoid is indefinite for $t^2 < 1$ and definite for $t^2 > 1$. The converse occurs in the hyperbolic case: if $e_2$ and $e_3$ have the same causal character, it is indefinite at any point. If they have distinct characters, it is definite for $t^2 < 1$ and Lorentz for $t^2 > 1$. In the parabolic case, we find that $\det g = -4|e_1|_p^2 t$, hence, if $|e_1|_p^2$ is positive, the surface is definite for negative $t$ and indefinite for positive $t$. This is the converse if $|e_1|_p^2$ is negative.

---

[4]This surface is called *Conjugate of Enneper's surface of the 2nd kind* in [Kobayashi (1983)] and [van de Woestijne (1990)].

**Remark 8.** All the helicoids are contained in a three-dimensional subspace of $\mathbb{R}^m$. The minimal cylinders and hyperbolic and parabolic helicoids occur only if the metric $\langle .,.\rangle_p$ is indefinite. In particular, the only non-planar, ruled, minimal surface in $(\mathbb{R}^m, \langle .,.\rangle_0)$ is the elliptic helicoid. In the case $m = 3$, this result is known as *Catalan theorem*, and is one of the first classification theorems about minimal surfaces. The classification of minimal, ruled surfaces of $(\mathbb{R}^3, \langle .,.\rangle_1)$ was given in [Kobayashi (1983)] in the spacelike case and in [van de Woestijne (1990)] in the indefinite case (see also [Niang (2003)]).

## 2.4 Weierstrass representation for minimal surfaces

Let $S$ be a non-degenerate surface in $(\mathbb{R}^m, \langle .,.\rangle_p)$. By Theorem 6, $S$ admits local isothermic coordinates $(s, t)$, i.e. it is the locally the image of an immersion $f$ of an open subset $U$ of $\mathbb{R}^2$, and we have

(1) $$|f_s|^2_p = \epsilon |f_t|^2_p$$

and

(2) $$\langle f_s, f_t \rangle_p = 0,$$

where $\epsilon = 1$ if the surface is definite, and $\epsilon = -1$ if it is indefinite. Differentiating these equations with respect to $s$ and $t$ and subtracting, we get

$(1)_s + (2)_t$ $$\langle f_{ss} + \epsilon f_{tt}, f_s \rangle_p = 0,$$

and

$(2)_s - (1)_t$ $$\langle f_{ss} + \epsilon f_{tt}, f_t \rangle_p = 0.$$

Hence the vector $f_{ss} + \epsilon f_{tt}$ is normal to $S$, and Equation (2.1) reads:

$$\vec{H} = \frac{1}{2} \frac{(f_{ss} + \epsilon f_{tt})^\perp}{|f_s|^2_p} = \frac{f_{ss} + \epsilon f_{tt}}{|f_s|^2_p}.$$

The last expression is nothing but the Laplacian of $f$ for the induced metric, expressed in the isothermic coordinates. Hence we have obtained the following:

**Theorem 9.** *The mean curvature vector of the immersion $f$ of a surface in $(\mathbb{R}^m, \langle .,.\rangle_p)$ is the Laplacian of $f$ for the induced metric. In particular, $f$ is minimal if and only if the coordinates functions are harmonic for the induced metric.*

In the definite case, remembering that a harmonic function on a compact, Riemannian manifold must be constant (Corollary 2, Chapter 1), we recover a particular case of Theorem 3:

**Proposition 6.** *There is no compact minimal surface with definite metric in* $(\mathbb{R}^m, \langle \cdot, \cdot \rangle_p)$.

The statement of Theorem 9 is coordinate-free, however we will go on using isothermic coordinates, since it will allow us to find explicit, local expressions for all minimal surfaces. In order to be more specific, we shall consider separately the two cases.

### 2.4.1   *The definite case*

Here $\epsilon = 1$, so in order to find minimal surfaces, we have to find $\mathbb{R}^m$-valued functions of the two variables $(s,t)$ which are harmonic. For this purpose we introduce the complex variable $z = s + it$, where $i := \sqrt{-1}$ and observe that the Laplace operator on $\mathbb{R}^2$ can be "factorized", i.e.

$$\Delta = \frac{\partial}{\partial s^2} + \frac{\partial}{\partial t^2} = \left( \frac{\partial}{\partial s} + i \frac{\partial}{\partial t} \right) \left( \frac{\partial}{\partial s} - i \frac{\partial}{\partial t} \right). \tag{2.9}$$

We introduce the *Cauchy-Riemann operators*

$$\frac{\partial}{\partial z} := \frac{1}{2} \left( \frac{\partial}{\partial s} - i \frac{\partial}{\partial t} \right) \quad \text{and} \quad \frac{\partial}{\partial \bar{z}} := \frac{1}{2} \left( \frac{\partial}{\partial s} + i \frac{\partial}{\partial t} \right),$$

and we say that a complex-valued function $\phi(s,t) = \phi(z)$ is *holomorphic* if

$$\frac{\partial \phi}{\partial \bar{z}} = 0.$$

It is well-known that a holomorphic function $\phi$ defined on a simply connected open subset of $\mathbb{C}$ admits a primitive, i.e. a function $\Phi$ such that $\frac{\partial \Phi}{\partial z} = \phi$, which is moreover unique up to an additive constant. As in the case of the primitive of a continuous function of the real variable, $\Phi$ may be defined as an integral: the assumption that $\phi$ is holomorphic insures that the one-form $\phi(z)dz$ is closed. It follows from Stokes theorem that the integral of $\phi(z)dz$ on a closed curve of a simply connected open subset of $\mathbb{C}$ is zero, and therefore that the integral on a path depends only on the end points of the path. It is then a simple exercise to check that $\Phi(z) = \int_z \phi(z)dz$, where the integral is meant to be over any curve whose end points are a fixed point[5] $z_0$ and the variable point $z$.

---

[5] Changing the fixed point $z_0$ amounts to change the additive constant.

It follows from Equation (2.9) that $\vec{H}$ vanishes if and only if the $m$ functions

$$\frac{\partial f}{\partial z} := (\phi_1, ..., \phi_m)$$

are holomorphic. Moreover we have $f(s,t) = \text{Re} \int_z \phi(z)dz$ and the conformality assumption on $f$ is equivalent to the algebraic equation

$$-\sum_{j=1}^{p} \phi_j^2 + \sum_{j=p+1}^{m} \phi_j^2 = 0. \qquad (2.10)$$

Finally, the facts that $f$ is an immersion and that the induced metric is non-degenerate are equivalent to the fact that

$$\left.\left|\frac{\partial f}{\partial x}\right|^2\right|_p = \left.\left|\frac{\partial f}{\partial y}\right|^2\right|_p = 2\left.\left|\frac{\partial f}{\partial z}\right|^2\right|_p \neq 0,$$

i.e.

$$-\sum_{j=1}^{p} |\phi_j|^2 + \sum_{j=p+1}^{m} |\phi_j|^2 \neq 0. \qquad (2.11)$$

Summing up, we have a recipe to build minimal surfaces with definite metric:

(i) Take a $\mathbb{C}^m$-valued holomorphic map (the *Weierstrass data*) $(\phi_1, ..., \phi_m)$ on an open subset $U$ of the complex plane satisfying the non-degeneracy condition (2.11) and the quadratic Equation (2.10);

(ii) Take $f(s,t) := \text{Re} \int_z \phi(z)dz$; then $f$ is a conformal, minimal immersion.

Given a solution $(\phi_1, ..., \phi_m)$ of (2.10), we see that $(i\phi_1, ..., i\phi_m)$ is another solution. The two minimal surfaces corresponding to the data $(\phi_1, ..., \phi_m)$ and $(i\phi_1, ..., i\phi_m)$ are said to be *conjugate*. Actually, from a Weierstrass data, i.e. $m$ holomorphic functions $\phi_1, ..., \phi_m$ satisfying Equation (2.10), we can construct not only a pair of conjugate minimal surfaces, but a whole one-parameter family, called an *associate* family: in the second point of the recipe above, set, for $\theta \in \mathbb{R}/2\pi\mathbb{Z}$,

$$f_\theta(s,t) := \cos\theta \, \text{Re} \int_z \phi(z)dz + \sin\theta \, \text{Im} \int_z \phi(z)dz \,.$$

It is easy to check that

$$\frac{\partial f_\theta}{\partial z} = e^{-i\theta}(\phi_1, ..., \phi_m).$$

Hence $\frac{\partial f_\theta}{\partial z}$ is holomorphic and still satisfies Equation (2.10) as well as the non-degeneracy condition (2.11). We conclude that $f_\theta$, for $\theta \in \mathbb{R}/2\pi\mathbb{Z}$ is a family of minimal immersions. Moreover, any two members of this family are isometric, i.e. for $\theta \neq \theta'$, the immersions $f_\theta$ and $f_{\theta'}$ induce the same metric on $U$. In other words, the intrinsic geometry of the two minimal surfaces $f_\theta(U)$ and $f_{\theta'}(U)$ is the same, although their extrinsic geometry is not *a priori*.

**Example 3. (Complex surfaces).** Let $\phi_1$ and $\phi_3$ two arbitrary holomorphic functions and set $\phi_2 = i\phi_1$ and $\phi_4 = i\phi_3$. Then we have both

$$\phi_1^2 + \phi_2^2 + \phi_3^2 + \phi_4^2 = 0$$

and

$$-\phi_1^2 - \phi_2^2 + \phi_3^2 + \phi_4^2 = 0,$$

hence we get a surface of $\mathbb{R}^4$ which is minimal for the two metrics $\langle .,.\rangle_0$ and $\langle .,.\rangle_2$. It is easily checked that the immersion $f := \mathrm{Re} \int_z (\phi_1, i\phi_1, \phi_3, i\phi_3)dz$ is holomorphic if we identify $\mathbb{R}^4$ with $\mathbb{C}^2$ setting ($z_1 = x_1 + ix_2, z_2 = x_3 + ix_4$). Such a surface is called *complex*, because its tangent planes are complex linear lines of $\mathbb{C}^2$. Complex submanifolds will be discussed in Chapter 5, Section 5.1 in a wider framework.

### 2.4.1.1    *The case of dimension* 3

Up to scaling by the transformation $g \mapsto -g$ (see Proposition 5 of Chapter 1), there are two flat pseudo-Riemannian metrics on $\mathbb{R}^3$: the Minkowski metric $\langle .,.\rangle_1$ and the Riemannian metric $\langle .,.\rangle_0$. It is easy to find explicit holomorphic functions $(\phi_1, \phi_2, \phi_3)$ such that $\phi_1^2 + \phi_2^2 - \phi_3^2 = 0$ or $\phi_1^2 + \phi_2^2 + \phi_3^2 = 0$:

**Theorem 10.**

(i) *(Classical Weierstrass representation) Let $F$ and $G$ be two holomorphic functions defined on a simply connected open subset $U$ of $\mathbb{C}$ such that $F$ does not vanish on $U$. Then the map*

$$f(s,t) = \mathrm{Re} \int_z \left( \frac{1}{2}F(1 - G^2), \frac{i}{2}F(1 + G^2), FG \right) dz$$

*is a minimal, conformal immersion of $U$ into $(\mathbb{R}^3, \langle .,.\rangle_0)$. Conversely, any minimal surface of $(\mathbb{R}^3, \langle .,.\rangle_0)$ may be locally parametrized by a conformal immersion of this form.*

(ii) *Let $F$ and $G$ be two holomorphic functions defined on a simply connected open subset $U$ of $\mathbb{C}$ such that $F$ does not vanish and $|G| \neq 1$ on $U$. Then the map*

$$f(s,t) = \text{Re} \int_z \left( \frac{1}{2} F(1 + G^2), \frac{i}{2} F(1 - G^2), FG \right) dz$$

*is a conformal immersion of $U$ into $(\mathbb{R}^3, \langle .,. \rangle_1)$ whose image is a minimal, spacelike surface. Conversely, any minimal, spacelike surface of $(\mathbb{R}^3, \langle .,. \rangle_1)$ may be locally parametrized by a conformal immersion of this form.*

*Proof.* The Riemannian case is the easier one and is left to the reader. In the Minkowski case, we check easily that $\phi_1^2 + \phi_2^2 - \phi_3^2$ vanishes and that

$$|\phi_1|^2 + |\phi_2|^2 - |\phi_3|^2 = \frac{|F|^2}{2}(1 - |G|^2)^2, \tag{2.12}$$

which is strictly positive as long as $F$ does not vanish and $|G| \neq 1$. Therefore $f(s,t)$ is a minimal, conformal immersion. Conversely, suppose $f$ is a conformal immersion of a simply connected open subset $U$ whose image is minimal and spacelike, and set $F := \phi_1 - i\phi_2$. Observe that we cannot have vanishing $F$: it would imply $\phi_1 = i\phi_2$, so by the quadric equation $\phi_1^2 + \phi_2^2 - \phi_3^2 = 0$ we would have vanishing $\phi_3$, and then the image of $f$ would be planar. Since $F$ does not vanish, we may set $G := \phi_3/F$. Therefore we have $(\phi_1, \phi_2, \phi_3) = (\frac{1}{2}F(1 + G^2), \frac{i}{2}F(1 - G^2), FG)$, and Equation (2.12) holds. Therefore we cannot have $|G| = 1$ by the non-degeneracy assumption (2.11). Hence the pair $(F, G)$ defined in this way is the Weierstrass data corresponding to the minimal immersion $f$. $\square$

**Example 4.** If we choose $F$ and $G$ constant, we obtain a linear map $f$, so its image is a plane. One of the simplest pairs one may think beyond the constants is $(F(z), G(z)) = (1, z^{-1})$. In the Minkowski case, we get

$$(\phi_1, \phi_2, \phi_3) = \left( \frac{i}{2}\left(1 - \frac{1}{z^2}\right), \frac{1}{2}\left(1 + \frac{1}{z^2}\right), \frac{1}{z} \right).$$

Rather than writing explicitly a path integral of $(\phi_1, \phi_2, \phi_3)$, we compute formally a primitive:

$$\left( \frac{i}{2}\left(z + \frac{1}{z}\right), \frac{1}{2}\left(z - \frac{1}{z}\right), \log z \right). \tag{2.13}$$

Introducing polar coordinates $z = re^{i\theta}$, the real part of the latter expression is:

$$f_0(r, \theta) = \left( -\frac{1}{2}\left(r - \frac{1}{r}\right)\sin\theta, -\frac{1}{2}\left(r - \frac{1}{r}\right)\cos\theta, \log r \right).$$

We observe that the image of $f_0$ is a surface of revolution: for fixed $r$, $\theta \mapsto f_0(r, \theta)$ is a horizontal circle. Thus we recover the spacelike catenoid of Section 2.2. The imaginary part of (2.13) is:

$$f_{\pi/2} = \left( \frac{1}{2}\left(r + \frac{1}{r}\right)\cos\theta, \frac{1}{2}\left(r + \frac{1}{r}\right)\sin\theta, \theta \right).$$

Restricting the parameter space to either $r < 1$ or $r > 1$, this is a local parametrization of the part of the elliptic helicoid which lies outside of the cylinder $\{x_1^2 + x_2^2 = 1\}$. Hence the spacelike catenoid and the helicoid are conjugate surfaces.

In the Riemannian case, the situation is very similar:

$$(\phi_1, \phi_2, \phi_3) = \left( \frac{1}{2}\left(1 - \frac{1}{z^2}\right), \frac{i}{2}\left(1 + \frac{1}{z^2}\right), \frac{1}{z} \right),$$

a primitive of which is

$$\left( \frac{1}{2}\left(z + \frac{1}{z}\right), \frac{i}{2}\left(z - \frac{1}{z}\right), \log z \right).$$

The real part of the latter expression gives, in polar coordinates:

$$f_0(r, \theta) = \left( \frac{1}{2}\left(r + \frac{1}{r}\right)\cos\theta, -\frac{1}{2}\left(r + \frac{1}{r}\right)\sin\theta, \log r \right).$$

The image of $f_0$ is the classical catenoid (Exercise 1). The imaginary part gives:

$$f_{\pi/2} = \left( \frac{1}{2}\left(r - \frac{1}{r}\right)\sin\theta, \frac{1}{2}\left(r - \frac{1}{r}\right)\cos\theta, \theta \right).$$

It is easily seen that it is a local parametrization of the elliptic helicoid.

### 2.4.2    *The indefinite case*

Here we have to find $\mathbb{R}^m$-valued maps $f$ such that $f_{ss} - f_{tt}$ and $|f_s|_p^2 + |f_t|_p^2$ vanish. As in the proof of Theorem 6, we introduce "null" coordinates $(u, v)$ for the induced metric, i.e. such that $\partial_u = \partial_s + \partial_t$ and $\partial_v = \partial_s - \partial_t$. Hence $f_{ss} - f_{tt} = f_{uv}$ and we have to solve $f_{uv} = 0$. This is maybe the simplest PDE of all: the general solution is $f(u, v) = \gamma_1(u) + \gamma_2(v)$, where $\gamma_1(u)$, $u \in I_1$ and $\gamma_2(v)$, $v \in I_2$ are two $\mathbb{R}^m$-valued functions. The assumption

$$|f_s|_p^2 + |f_t|_p^2 = 0$$

is equivalent to

$$|f_u|_p^2 = |\gamma_1(u)'|_p^2 = 0 \quad \text{and} \quad |f_v|_p^2 = |\gamma_2(v)'|_p^2 = 0,$$

so the velocities of the curves $\gamma_1$ and $\gamma_2$ are null vectors. Moreover, the induced metric is non-degenerate if and only if $\langle \gamma_1'(u), \gamma_2'(v) \rangle_p \neq 0, \forall (u,v) \in I_1 \times I_2$. Summing up, we have proved:

**Theorem 11.** *Let $S$ be an indefinite minimal surface of $(\mathbb{R}^m, \langle ., . \rangle_p)$. Then it may be locally parametrized by an immersion of the form*

$$f(u,v) = \gamma_1(u) + \gamma_2(v),$$

*where $\gamma_1, \gamma_2$ are two null curves of $\mathbb{R}^m$ such that $\langle \gamma_1'(u), \gamma_2'(v) \rangle_p \neq 0, \forall u, v$.*

**Remark 9.** This representation formula has been derived in [Weinstein (1996)] in the case of $(\mathbb{R}^3, \langle ., . \rangle_1)$ and in [Chen (2009)] for $(\mathbb{R}^4, \langle ., . \rangle_2)$. A different (although equivalent) point of view has been derived for the case of Minkowski space $(\mathbb{R}^3, \langle ., . \rangle_1)$ in [Magid (1991)]. This representation is more complicated and is restricted to the analytic case but resembles the definite case, since it involves a kind of analogs of holomorphic functions, called *split-holomorphic functions*.

**Example 5.** Assume that the null curve $\gamma_1$ is a straight line: $\gamma_1(u) = \gamma_0 u$, where $\gamma_0$ is a null direction. Then $f(u,v) = \gamma_0 u + \gamma_2(v)$: we recover the examples of cylindrical surfaces described in the previous section.

**Example 6. (The hyperbolic helicoid).** Consider the hyperbolic helicoid of Section 2.3, setting $c = 1$ for simplicity:

$$f(s,t) = (\cosh s\, e_1 + \sinh s e_2)t + e_3 s,$$

where $e_1$ is a unit negative vector and $e_2$ and $e_3$ are unit positive vectors. The parametrization is not isothermic, however it suffices to introduce the parameter $y$ such that $t = \sinh y$ to get such isothermic coordinates. Hence we get null coordinates $(u,v)$ setting $(u + v = s, u - v = y)$ and the parametrization becomes

$$f(u,v) = (\cosh(u+v)e_1 + \sinh(u+v)e_2)\sinh(u-v) + (u+v)e_3$$

$$= \frac{1}{2}(\sinh(2u) - \sinh(2v)e_1 + \frac{1}{2}(\cosh(2u) - \cosh(2v)e_2) + (u+v)e_3.$$

Hence, in this case, the two null curves are $\gamma_1(u) = \frac{1}{2}(\sinh(2u)e_1 + \cosh(2u)e_2) + ue_3$ and $\gamma_2(v) = -\frac{1}{2}(\sinh(2v)e_1 + \cosh(2v)e_2) + ve_3$.

We may check that

$$\langle \gamma_1'(u), \gamma_2'(v) \rangle_p = \cosh(2(u+v)) + 1 > 2,$$

so the immersion is always non-degenerate.

### 2.4.3   A remark on the regularity of minimal surfaces

The causal character of the induced metric of a minimal surface yields a striking contrast in the regularity of the surface: in the definite case, we show that a minimal surface is the image of the real part of a holomorphic function. The surface has therefore the highest degree of regularity: it is real analytic. On the contrary, the representation formula of minimal indefinite surfaces only requires the immersion to be twice differentiable.

## 2.5   Exercises

(1) Calculate the minimal surface equation for graph immersions $f(x_1, x_2) = (x_1, x_2, u(x_1, x_2))$ of $\mathbb{R}^3$ in the Riemannian case, i.e. the equivalent of Equation (2.2) when the metric is $\langle ., . \rangle_0 = dx_1^2 + dx_2^2 + dx_3^2$. Show that $u = \arg(x_1 + ix_2)$ is a solution of this equation, recovering the fact that the elliptic helicoid is also a minimal surface for the positive metric of $\mathbb{R}^3$. Show that a radial solution of this equation, i.e. of the form $u(x_1, x_2) := v(\sqrt{x_1^2 + x_2^2})$, must satisfy Equation (2.8) (it will become clear in the next chapter why this is the same equation). Draw a picture of the resulting surface. This is the classical *catenoid*.

(2) Seek several solutions of Equation (2.2) of the form $u(x_1, x_2) = F(x_1) + G(x_2)$, where $F$ and $G$ are functions of the real variable (Hint: two solutions of the equation $F'' = 1 - (F')^2$ are $F' = \frac{\sinh x}{\cosh x}$ and $F' = \frac{\cosh x}{\sinh x}$). Discuss the causal character of the corresponding surfaces. Using the same method, show that $u(x_2, x_3) = \log(\frac{\cos x_2}{\cosh x_3})$ is a solution of Equation (2.6) and draw it. We call it the *Lorentzian Scherk's surface*.

(3) Show that the spacelike hyperbolic catenoid and the spacelike portion of the hyperbolic helicoid are conjugate surfaces. They are therefore isometric.

(4) Show that the shape operator of the indefinite portion of the hyperbolic helicoid is not diagonalisable.

(5) (Enneper surfaces) Describe and draw the minimal surface obtained with the Weierstrass data $(F(z), G(z)) = (1, z)$, both in the Riemannian and Minkowski cases.

(6) Let $(\Sigma, g)$ be a Riemannian surface and $u$ a real function on $\Sigma$. The graph of $u$ is a surface in the product manifold $\mathcal{M} := \Sigma \times \mathbb{R}$, which we endow with the indefinite metric $\bar{g} := g - dt^2$, where $t$ is the canonical coordinate on $\mathbb{R}$. Discuss the causal character of the graph of $u$ in terms of the gradient of $u$ (recall the gradient depends on the metric $g$, see Section 1.1.3 of Chapter 1). Write the minimal surface equation for the graph of $u$ (hint: it is simpler to use isothermic coordinates on $(\Sigma, g)$).

# Chapter 3

# Equivariant minimal hypersurfaces in space forms

This chapter deals with submanifolds with co-dimension one, i.e. hypersurfaces. In the first section we introduce the space forms, which are the pseudo-Riemannian analogs of the Riemannian round sphere. We study both their extrinsic and intrinsic geometry: on the one hand we prove that they are, next to the hyperplanes, the simplest hypersurfaces of Euclidean space in terms of extrinsic curvature, and on the other hand we see that they are the simplest examples of non-flat pseudo-Riemannian manifolds. Equivariant hypersurfaces are, roughly speaking, those hypersurfaces which are the most symmetric ones next to space forms themselves. Sections 3.2 and 3.3 are devoted respectively to the description of minimal equivariant hypersurfaces of Euclidean space and of space forms.

## 3.1 The pseudo-Riemannian space forms

Consider the real space $\mathbb{R}^{n+1}, n + 1 \geq 3$, endowed with the pseudo-Riemannian metric of signature $(p, n + 1 - p)$:

$$\langle ., . \rangle_p := -\sum_{i=1}^{p} dx_i^2 + \sum_{i=p+1}^{n+1} dx_i^2.$$

The group of orientation preserving isometries (rotations) of $(\mathbb{R}^{n+1}, \langle ., . \rangle_p)$ is the group $SO(p, n + 1 - p)$ of invertible matrices of positive determinant which preserve the metric $\langle ., . \rangle_p$, i.e.

$$SO(p, n + 1 - p) := \left\{ M \in Gl(\mathbb{R}^{n+1}) \mid \langle MX, MY \rangle_p = \langle X, Y \rangle_p, \det M > 0 \right\}.$$

The orbits of $SO(p, n + 1 - p)$ are the quadrics

$$\mathbb{Q}_{p,c}^n := \left\{ x \in \mathbb{R}^{n+1} \mid |x|_p^2 = c \right\},$$

where $c$ is some real constant. They are smooth hypersurfaces, except when $c = 0$: the set $\mathbb{Q}^n_{p,0}$, if non-empty, has a unique singularity at $0$ and is actually a cone. It is usually called *the light cone*, which has a physical meaning in the Minkowski case. The induced metric is degenerate on the light cone, therefore no curvature can be defined on it and in the following, we shall always deal with the case $c \neq 0$.

The next theorem makes precise our claim that the hypersurfaces $\mathbb{Q}^n_{p,c}$ are, beyond the hyperplanes, the simplest ones of $\mathbb{R}^{n+1}$:

**Theorem 12.** *The quadric $\mathbb{Q}^n_{p,c}$ is a totally umbilic hypersurface of $\mathbb{R}^{n+1}$ and has constant mean curvature $\kappa := |c|^{-1/2}$ with respect to the unit normal vector field $N(x) := -\kappa x$. Conversely, if $S$ is a connected, totally umbilic hypersurface of $\mathbb{R}^{n+1}$, its mean curvature curvature $\kappa$ is constant and $S$ is either an open subset of a hyperplane (if $\kappa$ vanishes) or, up to translation, one of the quadric $\mathbb{Q}^n_{p,c}$, with $|c| = 1/\kappa^2$.*

*Proof.* Let $X$ be some tangent vector to $\mathbb{Q}^n_{p,c}$ at some point $x$ and $x(s)$ a regular curve in $\mathbb{Q}^n_{p,c}$ such that $x(0) = x$ and $x'(0) = X$. Differentiating the identity $\langle x(s), x(s) \rangle_p = c$ at $s = 0$, we find $2 \langle X, x \rangle_p = 0$, so the tangent space to $\mathbb{Q}^n_{p,c}$ at $x$ is $x^{\perp}$. Therefore $N_x \mathbb{Q}^n_{p,c} = x \mathbb{R}$. Setting $N := -|c|^{-1/2} x = -\kappa x$, we calculate $|N|^2_p = |c|^{-1} |x|^2_p = \frac{c}{|c|}$, hence $N$ is a unit normal vector. Next we compute the shape operator:

$$D_X N = -|c|^{-1/2} D_X x = -\kappa X,$$

so that $A_N = \kappa Id$ and $\mathbb{Q}^n_{p,c}$ is totally umbilic as claimed.

Conversely, let $S$ a totally umbilic hypersurface of $\mathbb{R}^{n+1}$ with a local unit normal vector field $N$: there exists a function $\kappa(x)$ on $S$ such that $D_X N = -\kappa X$, where $X$ is any tangent vector field. We first prove that if $S$ is totally umbilic, then $\kappa$ does not depend on the point $x$: differentiating the identities $D_X N = -\kappa(x) X$ and $D_Y N = -\kappa(x) Y$ with respect to $Y$ and $X$ respectively, we get

$$D_Y D_X N = -Y(\kappa) X - \kappa D_Y X$$

and

$$D_X D_Y N = -X(\kappa) Y - \kappa D_X Y.$$

Subtracting these two equations, we obtain

$$D_Y D_X N - D_X D_Y N = X(\kappa) Y - Y(\kappa) X - \kappa [X, Y].$$

Since $D_{[X,Y]} N = -\kappa [X, Y]$, we get

$$D_Y D_X N - D_X D_Y N - D_{[X,Y]} N = X(\kappa) Y - Y(\kappa) X$$

and

$$R^D(X,Y)Z = X(\kappa)Y - Y(\kappa)X.$$

The curvature $R^D$ of $D$ being zero, the right hand side term above vanishes. Hence, choosing $X$ and $Y$ not collinear, we get that both $X(\kappa)$ and $Y(\kappa)$ vanish. Since $\mathcal{S}$ is assumed to be connected, $\kappa$ is constant.

The totally geodesic case, i.e. of vanishing $\kappa$, has already been treated in Chapter 1, Proposition 4. However, the co-dimension one assumption allows a simpler argument here: since $D_X N = 0$ and $\mathcal{S}$ is connected, $N$ is a constant vector. Moreover, choosing any point $x_0$ of $\mathcal{S}$, we have $D_X \langle x - x_0, N \rangle_p = \langle X, N \rangle_p = 0$. Hence $\mathcal{S}$ is contained in the hyperplane $x_0 + N^\perp$.

If $\kappa$ does not vanish, we first prove that $x + N(x)/\kappa$ is constant: since $\kappa$ itself is constant, we have:

$$D_X(x + N/\kappa) = X + \frac{D_X N}{\kappa} = 0.$$

Hence, the point $x_0 := x + N/\kappa$ is constant (this is the "center" of $\mathcal{S}$). It follows that

$$\langle x - x_0, x - x_0 \rangle_p = \frac{|N|_p^2}{\kappa^2}.$$

Setting $c := \frac{|N|_p^2}{\kappa^2}$, we conclude that $\mathcal{S} \subset \mathbb{Q}_{p,c}^n + x_0$. In the positive case, i.e. $p = 0$, $\mathcal{S}$ is the sphere of radius $\sqrt{c}$ centered at $x_0$. $\qquad\square$

The next result shows that the intrinsic curvature of $\mathbb{Q}_{p,c}^n$ is also very simple.

**Proposition 7.** *The curvature tensor $R^\nabla$ of the induced metric on $\mathbb{Q}_{p,c}^n$ is given by the formula*

$$R^\nabla(X,Y)Z = \frac{1}{c}\big(\langle X, Z\rangle_p Y - \langle Y, Z\rangle_p X\big). \qquad (3.1)$$

*Moreover $\mathbb{Q}_{p,c}^n$ has constant sectional curvature $1/c$.*

**Remark 10.** One can prove that conversely, a pseudo-Riemannian manifold of constant sectional curvature is locally homeomorphic to one of the space forms $\mathbb{Q}_{p,c}^n$. A proof of this fact can be found in [Kriele (1999)].

*Proof of Proposition 7.* By the totally umbilicity of $\mathbb{Q}_{p,c}^n$, we have $h(X,Y) = \kappa\langle X, Y\rangle_p$ (see Section 1.2.4, Chapter 1), so we obtain the following expression for the Gauss formula:

$$D_X Y = \nabla_X Y + \epsilon\kappa\langle X, Y\rangle_p N.$$

Next choosing three tangent vector fields $X$, $Y$ and $Z$ on $\mathbb{Q}^n_{p,c}$, we calculate:

$$R^D(X,Y)Z = D_Y D_X Z - D_X D_Y Z + D_{[X,Y]}Z$$
$$= D_Y(\nabla_X Z + \epsilon\kappa\langle X,Z\rangle_p N) - D_X(\nabla_Y Z + \epsilon\kappa\langle Y,Z\rangle_p N)$$
$$+ \nabla_{[X,Y]}Z + \epsilon\kappa\langle[X,Y],Z\rangle_p N$$
$$= \nabla_Y\nabla_X Z + \epsilon\kappa\langle Y,\nabla_X Z\rangle_p N + \epsilon\kappa Y(\langle X,Z\rangle_p)N - \epsilon\kappa^2\langle X,Z\rangle_p Y$$
$$- \nabla_X\nabla_Y Z + \epsilon\kappa\langle X,\nabla_Y Z\rangle_p N + \epsilon\kappa X(\langle Y,Z\rangle_p)N$$
$$- \epsilon\kappa^2\langle Y,Z\rangle_p X + \nabla_{[X,Y]}Z + \epsilon\kappa\langle[X,Y],Z\rangle_p N.$$

Taking the tangent part of this expression, we get

$$0 = \nabla_Y\nabla_X Z - \nabla_X\nabla_Y Z + \nabla_{[X,Y]}Z - \epsilon\kappa^2(\langle X,Z\rangle_p Y - \langle Y,Z\rangle_p X),$$

so

$$R^\nabla = \nabla_Y\nabla_X Z - \nabla_X\nabla_Y Z + \nabla_{[X,Y]}Z$$
$$= \epsilon\kappa^2(\langle X,Z\rangle_p Y - \langle Y,Z\rangle_p X),$$

which is the required formula for $R\nabla$ since $\epsilon\kappa^2 = c^{-1}$.

Finally, we compute

$$\langle R^\nabla(X,Y)X,Y\rangle_p = \frac{1}{c}(|X|^2_p|Y|^2_p - \langle X,Y\rangle^2_p)$$

and we conclude that the sectional curvature of the plane $P$ spanned by $X$ and $Y$ is

$$K(P) = \frac{\langle R^\nabla(X,Y)X,Y\rangle_p}{|X|^2_p|Y|^2_p - \langle X,Y\rangle^2_p} = \frac{1}{c}. \qquad \square$$

We now claim that regarding the space form $\mathbb{Q}^n_{p,c}$ as an abstract pseudo-Riemannian manifold (this will be the subject of Section 3.3), one can first reduce the study to the case $c = 1$: firstly, the scaling transformation $\phi$ of $\mathbb{R}^{n+1}$ defined by $\phi(x) = \frac{1}{r}.x$ maps $\mathbb{Q}^n_{p,c}$ diffeomorphically onto $\mathbb{Q}^n_{p,\epsilon}$, and, for a pair of tangent vectors $X,Y$ of $\mathbb{Q}^n_{p,c}$, $\langle d\phi(X), d\phi(Y)\rangle_p = \frac{1}{r^2}\langle X,Y\rangle_p$. Hence, regarded as abstract manifolds, $\mathbb{Q}^n_{p,c}$ and $\mathbb{Q}^n_{p,\epsilon}$ are the same one with two different metrics, which are proportional. Therefore, by Proposition 5 of Chapter 1, they have the same minimal submanifolds. Moreover the sets $\mathbb{Q}^n_{p,-1}$ and $\mathbb{Q}^n_{n+1-p,1}$ are *anti-isometric:* they are diffeomorphic by the transformation $\tau : (x_1,...,x_p,x_{p+1},...,x_n) \mapsto (x_{p+1},...,x_n,x_1,...,x_p)$ and $\langle d\tau(X), d\tau(Y)\rangle_{n+1-p} = -\langle X,Y\rangle_p$. Again the two metrics are proportional and the two quadrics have the same pseudo-Riemannian geometry. We conclude that there are essentially $n + 1$ space forms of dimension $n$, the quadrics $\mathbb{Q}^n_p$, with $0 \leq p \leq n$. Some of them are well-known Riemannian or pseudo-Riemannian manifolds:

(i) $\mathbb{S}^n := \mathbb{Q}_{0,1}^n$ is the classical, round sphere; this is the only compact space form; the induced metric is definite positive.

(ii) $\mathbb{Q}_{n,1}^n$ is the only space form which is not connected. More precisely, it has two isometric connected components. For this reason, we set the *hyperbolic space* (or more precisely the *hyperboloid model* of the hyperbolic space) to be $\mathbb{H}^n := \mathbb{Q}_{n,1}^n \cap \{x \in \mathbb{R}^{n+1} | x_1 > 0\}$. The induced metric is definite negative. As this may cause some discomfort, the following definition, obviously equivalent, but which makes the metric positive, is usually preferred when dealing with Riemannian geometry: $\mathbb{H}^n := \mathbb{Q}_{1,-1}^n \cap \{x \in \mathbb{R}^{n+1} | x_{n+1} > 0\}$.

(iii) $d\mathbb{S}^n := \mathbb{Q}_{1,1}^n$ is called the *de Sitter space*; the induced metric is Lorentzian;

(iv) $Ad\mathbb{S}^n := \mathbb{Q}_{n-1,1}^n$ is called the *anti de Sitter space*[1]; the induced metric is, up to an anti-isometry, Lorentzian as well.

**Remark 11.** When $n = 1$, there are of course merely two "curve forms": the circle $\mathbb{S}^1 := \{x_1^2 + x_2^2 = 1\}$ and the hyperbola $\mathbb{H}^1 = \{-x_1^2 + x_2^2 = 1\}$. When $n = 2$, there are three "surface forms": the sphere $\mathbb{S}^2$ and the hyperbolic plane $\mathbb{H}^2$ have definite metric, while the Sitter and the anti de Sitter surfaces are the same one, with signature $(1,1)$. The four-dimensional Sitter and anti de Sitter spaces $d\mathbb{S}^4$ and $Ad\mathbb{S}^4$ are relevant in Relativity theory: the three positive directions represent the physical space, while the negative one represents time (see [Kriele (1999)]).

## 3.2 Equivariant minimal hypersurfaces in pseudo-Euclidean space

### 3.2.1 *Equivariant hypersurfaces in pseudo-Euclidean space*

As we have already stated, the group $SO(p, n + 1 - p)$ of rotations of $(\mathbb{R}^{n+1}, \langle ., . \rangle_p)$ is the group of matrices $M$ of positive determinant which preserve the metric $\langle ., . \rangle_p$, i.e. such that $\langle Mx, My \rangle_p = \langle x, y \rangle_p$. We have furthermore discussed those non-degenerate hypersurfaces of $\mathbb{R}^{n+1}$ which are invariant by the action of $SO(p, n + 1 - p)$ and seen that they have constant, non-vanishing mean curvature. In particular, none of them is minimal. In the following, we shall look for hypersurfaces of the Euclidean space which are minimal and invariant by the action of the following

---

[1]The usual definition is $Ad\mathbb{S}^n := \mathbb{Q}_{2,-1}^n$.

subgroup of $SO(p, n+1-p)$:

$$\widetilde{SO}(p, n-p) := \left\{ \begin{pmatrix} M & 0 \\ 0 & 1 \end{pmatrix} \Big| M \in SO(p, n-p) \right\}.$$

This subgroup is very important for the following reason: let us fix a positive point $x_0$ of $\mathbb{R}^{n+1}$. By linearity, if a rotation $M$ fixes $x_0$, i.e. $Mx_0 = x_0$, then it fixes also the point $e := x_0.\langle x_0, x_0 \rangle^{-1/2}$ of $\mathbb{Q}^n_{p,1}$. Since the action of $SO(p, n+1-p)$ is transitive on $\mathbb{Q}^n_{p,1}$, there exists a rotation $M_0$ that maps the "north pole" $(0, ..., 0, 1)$ of $\mathbb{Q}^n_{p,1}$ on $e$, i.e. $M_0(0, ..., 0, 1) = e$. Hence the subgroup of rotations of $(\mathbb{R}^{n+1}, \langle ., . \rangle_p)$ that fix $x_0$ is exactly

$$M_0^{-1}.\widetilde{SO}(p, n-p).M_0 := \{M_0^{-1}.M.M_0 \big| M \in \widetilde{SO}(p, n-p)\}.$$

This subgroup is conjugate to $\widetilde{SO}(p, n-p)$. Analogously, when $p \geq 1$, the group of rotations that fix a negative direction of $\mathbb{R}^{n+1}$ is conjugate to the group

$$\widetilde{SO}(p-1, n+1-p) := \left\{ \begin{pmatrix} 1 & 0 \\ 0 & M \end{pmatrix} \Big| M \in SO(p-1, n+1-p) \right\}.$$

Another interesting point for us is that the non-degenerate orbits of these groups are $(n-1)$-dimensional: let $\bar{x} = (x_0, z)$ be a point of $\mathbb{R}^{n+1}$, where $x_0 = (x_1, ..., x_n) \in \mathbb{R}^n$, and set $c := |x_0|^2_p$. It is not difficult to see that the orbit of $\bar{x}$ by the action of $\widetilde{SO}(p, n-p)$ is the set:

$$(\mathbb{Q}^{n-1}_{p,c}, z) := \left\{ (x, z) \in \mathbb{R}^{n+1} \big| x \in \mathbb{Q}^{n-1}_{p,c} \right\}.$$

Hence an $SO(p, n-p)$-equivariant hypersurface $\mathcal{S}$ of $\mathbb{R}^{n+1}$ is the union of a one-parameter family of submanifolds of the form $(\mathbb{Q}^{n-1}_{p,c(s)}, z(s))$, where $c$ and $z$ are two real functions defined on some interval $I$. We claim that if $c$ vanish on a sub-interval $J$ of $I$, then the hypersurface $f(J \times \mathbb{Q}^{n-1}_{p,0})$ is degenerate (this will be proved rigorously in the next section). Therefore, since our discussion is local, we assume from now on that $c$ does not vanish and we set $\epsilon_0 := c/|c|$, which is therefore constant, and $r := \sqrt{|c|}$, which is strictly positive. Finally $\mathcal{S}$ may be locally parametrized by an immersion of the form

$$\begin{aligned} f : I \times \mathbb{Q}^{n-1}_{p,\epsilon_0} &\to \mathbb{R}^{n+1} \\ (s, x) &\mapsto (r(s)\iota(x), z(s)), \end{aligned}$$

where $\iota : \mathbb{Q}^{n-1}_{p,\epsilon_0} \to \mathbb{R}^n$ denotes the canonical immersion of $\mathbb{Q}^{n-1}_{p,\epsilon_0}$, regarded as an abstract manifold, into $\mathbb{R}^n$, and $\gamma(s) := (r(s), z(s))$ is a parametrized curve in the half-plane $P := (0, \infty) \times \mathbb{R}$. Analogously, an $SO(p-1, n+1-p)$-equivariant hypersurface may be locally parametrized by an immersion of the form

$$g : I \times \mathbb{Q}^{n-1}_{p-1, \epsilon_0} \to \mathbb{R}^{n+1}$$
$$(s, x) \mapsto (z(s), r(s)\iota(x)).$$

In the next section we shall characterize equivariant minimal hypersurfaces of $\mathbb{R}^{n+1}$.

### 3.2.2 The minimal equation

In order to treat simultaneously the two cases described above, we make some slight changes in the notation. First, in the $SO(p-1, n+1-p)$ case, we re-arrange the order of the coordinates as follows $(x_1, ..., x_{n+1}) \mapsto (x_2, ..., x_{n+1}, x_1)$. Observe that with this new notation, the last direction is not necessarily positive. Therefore we set $\epsilon := |(0, ..., 0, 1)|_p^2$. We also set $p' := p$ in the first case and $p' := p - 1$ in the second one, so that

$$\langle ., . \rangle_p := \langle ., . \rangle_{p'} + \epsilon dx_{n+1}^2.$$

Finally, we claim that there is no loss of generality in assuming that $\epsilon_0 = 1$: otherwise, replace the metric $\langle ., . \rangle_p$ by $\langle ., . \rangle_{n+1-p}$ and re-arrange the order of the coordinates as follows: $(x_1, ..., x_p, x_{p+1}, ..., x_{n+1}) \mapsto (x_{p+1}, ..., x_{n+1}, x_1, ..., x_p)$. Hence we are left to study the geometry of the immersion.

$$f : I \times \mathbb{Q}^{n-1}_{p', 1} \to \mathbb{R}^{n+1}$$
$$(s, x) \mapsto (r(s)\iota(x), z(s)).$$

We introduce local coordinates $(s_1, ..., s_{n-1})$ in $\mathbb{Q}^{n-1}_{p', 1}$, so that $(s_0 := s, s_1, ..., s_{n-1})$ are local coordinates on $I \times \mathbb{Q}^{n-1}_{p', 1}$. We set $X_i := d\iota(\partial_{s_i})$, so in particular $(X_1, ..., X_{n-1})$ is a frame of tangent vectors along $\mathbb{Q}^{n-1}_{p', 1}$. We shall frequently write $x$ instead of $\iota(x)$, as it should not create confusion.

Computing the first derivatives of $f$,

$$f_s = (r'x, z') \quad \text{and} \quad df(\partial_{s_i}) = (rX_i, 0),$$

we see that $f$ is an immersion if and only if $(r', z')$ do not vanish and $r > 0$. In particular, the curve $\gamma$ is regular. We claim that the geometry of the immersion $f$ is described by the geometry of $\gamma$ in $P$ endowed with the pseudo-Riemannian metric $\langle ., . \rangle_{p''} = dr^2 + \epsilon dz^2$, where $p'' = p - p'$: denoting by $g$ the metric on $\mathbb{Q}^{n-1}_{p, 1}$ the coefficients of the induced metric $\bar{g}$ on $I \times \mathbb{Q}^{n-1}_{p, 1}$ are:

$$\bar{g}_{00} = (r')^2 + \epsilon(z')^2 = |\gamma'|_{p''}^2, \quad \bar{g}_{0i} = 0 \quad \text{and} \quad \bar{g}_{ij} = r^2 g_{ij}.$$

We may write these relations in a more compact, coordinate free form:

$$\bar{g} = |\gamma'|^2_{p''} ds^2 + r^2 g.$$

It follows that the immersion is degenerate if and only if $\bar{g}_{00}$ vanishes, i.e. $\gamma$ is degenerate (null). Hence we assume from now on that $\gamma$ is a non-null curve and that $s$ is the arc length parameter of the curve i.e. $|\gamma'|^2_{p''} = \epsilon'$, where $\epsilon' = 1$ or $-1$. We set $\nu := (\nu_1, \nu_2) = (z', -\epsilon r')$; it is one of the two unit normal vectors of $\gamma$ in $P$. We recall that the Frénet equations of $\gamma$ (see Chapter 1, Section 1.2.3.3) are

$$\begin{cases} \gamma'' = \epsilon \kappa \nu \\ \nu' = -\kappa \gamma'. \end{cases}$$

We now proceed to compute the second fundamental form of $f$. Observe that a unit normal vector to $f$ is given by $N = (\nu_1 x, \nu_2)$. Thus, using the Frénet equations, we get

$$D_{\partial_s} N = (\nu_1' \iota(x), \nu_2') = -\kappa(r'x, z') = -\kappa f_s,$$
$$D_{\partial_{s_i}} N = (\nu_1 X_i, 0).$$

Hence the coefficients of the second fundamental form $\bar{h}$ of $f$ are

$$\bar{h}_{00} = -\bar{g}\left(D_{\partial_s} N, \partial_s\right) = \kappa |f_s|^2_p = \kappa \bar{g}_{00},$$
$$\bar{h}_{0i} = -\bar{g}\left(D_{\partial_{s_i}} N, \partial_s\right) = -\langle \nu_1 X_i, r'x \rangle_{p'} = 0,$$
$$\bar{h}_{ij} = -\bar{g}\left(D_{\partial_{s_i}} N, \partial_{s_j}\right) = -\langle \nu_1 X_i, X_j \rangle_{p'} = -\nu_1 r \bar{g}_{ij}.$$

In particular, the shape operator of $f$ is diagonalizable and its principal curvatures are $\kappa_0 = \kappa$ with multiplicity one and $\kappa_1 = -\frac{\nu_1}{r}$ with multiplicity $n - 1$; it follows that the mean curvature of $f$ is

$$H = \frac{1}{n}\left(\kappa - (n-1)\frac{\nu_1}{r}\right). \tag{3.2}$$

In order to find at which condition the immersion is minimal, we must study the equation

$$\kappa - (n-1)\frac{\nu_1}{r} = 0 \tag{3.3}$$

for $\gamma$. It is a second order ODE, which therefore admits local solutions. In order to understand the global behaviour of these solutions, we shall look for a *first integral*, i.e. a function $E(r, z, r', z') = E(\gamma, \gamma')$ of $\gamma$ and its derivative such that if $\gamma(s)$ is any solution of (3.3), then $s \mapsto E(\gamma(s), \gamma'(s))$ is constant. It follows that the integral curves of the equation are contained in the level sets of $E$ and therefore may be described more easily.

We claim that $E(\gamma, \gamma') := \nu_1 r^{n-1}$ is a first integral of Equation (3.3) (observe that $\nu_1 = -\epsilon z'$ depends on $\gamma'$): by the Frénet equation $\nu_1' = -\kappa r'$, we have

$$\frac{d}{ds}E = \nu_1' r^{n-1} + (n-1)\nu_1 r' r^{(n-1)-1}$$
$$= r' r^{n-1}\left(-\kappa + (n-1)\frac{\nu_1}{r}\right),$$

which vanishes precisely if (3.3) holds.

We can discover a very simple solution of (3.3) assuming that $E$ vanishes. Since $r > 0$, it implies that $\nu_1$ vanishes, hence the normal vector is horizontal, and $\gamma$ is simply a horizontal straight line. The corresponding hypersurface is the horizontal hyperplane $\{x_{n+1} = z_0\}$ of $\mathbb{R}^{n+1}$, where $z_0$ is a real constant, which is totally geodesic.

Observe moreover that $E$ does not depend on the variable $z$. On the other hand, the derivative $\gamma'$ is completely determined by $\nu_1 = -\epsilon z'$, since $r' = \sqrt{1 - \epsilon(z')^2}$. These two crucial facts imply that to have found $E$ is sufficient to get a fine description of the integral curves. In classical terms, we would say that we have integrated Equation (3.3) "by quadratures." It turns out that the features of the curves $\gamma$ depend on the different values of $\epsilon$ and $\epsilon'$, so it is convenient to treat these different cases separately. This will be the object of the next sub-sections.

### 3.2.3 *The definite case* $(\epsilon, \epsilon') = (1, 1)$

By the arc length parametrization assumption, $\gamma' = (r', z') = (\cos\theta, \sin\theta)$, where $\theta$ denotes the angle made by the tangent vector $\gamma$ with the horizontal direction. It follows that $\nu_1 = -\sin\theta$, and

$$E(\gamma, \gamma') = E(r, \theta) = -\sin\theta\, r^{n-1}.$$

Hence the projections of the solutions $(\gamma, \gamma')$ onto the plane $(r, \theta)$ are the graphs $(r(\theta), \theta) = ((\frac{-E}{\sin\theta})^{1/(n-1)}, \theta)$. These curves are unbounded, and moreover contained in the horizontal strip $\{k\pi < \theta < (k+1)\pi\}$. Without loss of generality, we may assume that $k = 0$, so that $\theta \in (0, \pi)$; this implies that $E < 0$. The coordinate $r$ has a unique minimum $r_0 := (-E)^{1/(n-1)}$ at $\theta = \pi/2$ and they are symmetric with respect to the horizontal line $\{\theta = \pi/2\}$. To get a picture of the curve $\gamma$ and hence of the minimal immersion $f$, it remains to express the function $z$ in terms of $r$. Without loss of generality, we set $r(0) = r_0$ and $z(0) = 0$, so that:

$$z(s) = \int_0^s z'(\sigma)d\sigma = \int_0^s \sin\theta(\sigma)d\sigma = \int_0^s \frac{r^{n-1}(\sigma)}{(-E)}d\sigma.$$

On the upper branch of the curve, i.e. for positive $z$, we may use $r$ as a variable. By the definition of $\theta$, $dr = \cos\theta ds$, hence, taking into account that $\cos\theta > 0$,

$$z_{r_0}(r) = \int_{r_0}^r \frac{\sin\theta}{\cos\theta} d\rho = \int_{r_0}^r \left( \left( \frac{\rho^{n-1}}{(-E)} \right)^2 - 1 \right)^{-1/2} d\rho$$

$$= \int_{r_0}^r \left( \left( \frac{\rho}{r_0} \right)^{2(n-1)} - 1 \right)^{-1/2} d\rho.$$

We claim that varying the constant $r_0$ leads to different but homothetic curves: by a simple change of variable, we see that $z_{r_0}(r/r_0) = \frac{1}{r_0} z_1(r)$. It follows that the curve $r \mapsto (r, z_{r_0}(r))$ is the homothetic of the curve $r \mapsto z_1(r)$.

We conclude with the description of the asymptotic behaviour of $\gamma$:

$$z(r) \simeq \int_r \rho^{-1/(n-1)} d\rho,$$

hence the situation depends on the dimension $n$: if $n = 2$, $\lim_{r\to\infty} z = \pm\infty$, while if $n \geq 2$, $\sup_\gamma |z| < \infty$, i.e. the curve $\gamma$ is contained in a horizontal strip $\{(r, z) \in P, |z| < z_0\}$, where $z_0$ is real constant.

**Remark 12.** In the case $n = 2$, we have $z(r) = r_0 \cosh^{-1}(r/r_0)$ and we recover two surfaces described in Chapter 2: if $p = 0$, it is the classical catenoid (see Example 4 and Exercise 1) while if $p = 1$, it is the case of the Lorentzian hyperbolic catenoid of the first kind (Section 2). For $n \geq 3$, we may obtain explicit, but very complicated expressions of the function $z(r)$ in terms of elliptic functions[2], but there are not necessary since we already derived the geometric features of the curve $\gamma$.

### 3.2.4   *The indefinite positive case* $(\epsilon, \epsilon') = (-1, 1)$

Since the metric is indefinite and $(r')^2 - (z')^2 = 1$, it is convenient to introduce the "hyperbolic angle" $\theta$ made by the tangent vector $\gamma$ with the horizontal direction, i.e. the real function $\theta$ uniquely determined by $(r', z') = (\cosh\theta, \sinh\theta)$. It follows that $\nu_1 = \sinh\theta$ and

$$E = E(r, \theta) = \sinh\theta \, r^{n-1}.$$

We deduce that the projections $(r, \theta)$ of the integral curves are contained in one of the quadrants $\{\theta > 0\}$ and $\{\theta < 0\}$ of $P$. Moreover, since

$$\frac{dz}{dr} = \frac{\sinh\theta}{\cosh\theta} = \left( \frac{r^{2(n-1)}}{E^2} + 1 \right)^{-1/2},$$

---

[2]For example using the on line integrator http://integrals.wolfram.com/index.jsp

the integral $\int \frac{dz}{dr}(\rho)d\rho$ is convergent at $r = 0$. So we may assume that $r(0) = 0$ and $z(0) = 0$, and we obtain

$$z(r) = \int_0^r \left(\frac{\rho^{2(n-1)}}{E^2} + 1\right)^{-1/2} d\rho.$$

We easily check that $\lim_{r\to\infty} z = \infty$ for $n = 2$ and that $\lim_{r\to\infty} < \infty$ if $n \neq 3$. Moreover the curve $\gamma$ is asymptotically null at $r = 0$.

**Remark 13.** In the case $n = 2$, we get $z(r) = r_0 \sinh^{-1}(r/r_0)$ and we recover two more surfaces described in Chapter 2: for $p = 1$, we fall back to the case of the Lorentzian hyperbolic catenoid of the second kind, while if $p = 2$, we get the spacelike elliptic catenoid. For $n \geq 3$, we may again express $z(r)$ in terms of elliptic functions.

### 3.2.5 *The indefinite negative case* $(\epsilon, \epsilon') = (-1, -1)$

Since $(r')^2 - (z')^2 = 1$, the "hyperbolic angle" $\theta$ here is the real function $\theta$ uniquely determined by $(r', z') = (\sinh\theta, \cosh\theta)$. It follows that $\nu_1 = \cosh\theta$ and

$$E(\gamma, \gamma') = E(r, \theta) = \cosh\theta\, r^{n-1}.$$

We deduce that the projections $(r, \theta)$ of the integral curves are symmetric with respect to the line $\{\theta = 0\}$ and that $r$ has a unique maximum $r_0 :=$ $|E|^{1/(n-1)}$ at $\theta = 0$. The curves $\gamma$ are also symmetric with respect to the line $\{z = 0\}$ and their upper branch corresponds to $\theta \leq 0$. In this case we have, using the fact that $\sinh\theta = -\sqrt{\cosh^2\theta - 1}$, we have

$$\frac{dz}{dr} = \frac{\cosh\theta}{\sinh\theta} = -\left(1 - \left(\frac{r^{n-1}}{E}\right)^2\right)^{-1/2}$$

$$= -\left(1 - \left(\frac{r}{r_0}\right)^{2(n-1)}\right)^{-1/2}.$$

On the other hand $\frac{dz}{dr} \simeq_{r\simeq r_0} -(r_0 - r)^{-1/2}$, so the integral $\int \frac{dz}{dr}(\rho)d\rho$ is convergent at $r_0$. Setting $r(0) = r_0$ and $z(0) = 0$, we deduce:

$$z(r) = -\int_{r_0}^r \frac{\cosh\theta}{\sinh\theta}d\rho = \int_r^{r_0} \left(1 - \left(\frac{\rho^{n-1}}{E}\right)^2\right)^{-1/2} d\rho$$

$$= \int_r^{r_0} \left(1 - \left(\frac{\rho}{r_0}\right)^{2(n-1)}\right)^{-1/2} d\rho.$$

We observe that $\lim_{r \to 0} |z| := z_0 < \infty$ so the curve $\gamma$ is bounded. Moreover, the tangents of $\gamma$ at the two end points $(0, z_0)$ and $(0, -z_0)$ become asymptotically null, and the corresponding leaves $\mathbb{Q}_{p,-r}^{n-1}$ become degenerate at the limit. As the arc length tends to infinity, the curve, although bounded, has infinite length and the corresponding hypersurface is complete. In the case $p = n$, the compact leaves $\mathbb{Q}_{n-1,-1}^{n-1} \simeq \mathbb{S}^{n-1}$ degenerate to points, so the image of the immersion $f$ is homeomorphic to the $n$-dimensional sphere with two punctures. In the other cases, the leaves degenerate to two light cones when $r \to 0$, so the corresponding hypersurface is an open cylinder $(-z_0, z_0) \times \mathbb{Q}_{p,-1}^{n-1}$.

**Remark 14.** In the case $n = 2$, we get $z(r) = r_0 \sin^{-1}(r/r_0)$: if $p = 1$, it is the spacelike hyperbolic catenoid; if $p = 2$, the metric we are considering is $-dx_1^2 - dx_2^2 + dx_3^2$, which is anti-isometric to the Minkowski metric. In this case we get the Lorentzian elliptic catenoid since the fiber are circles $\{-x_1^2 - x_2^2 = -r^2\}$.

### 3.2.6    Conclusion

We can summarize the results of this section as follows:

**Theorem 13.** *Let $S$ be a connected, minimal hypersurface of pseudo-Euclidean space $(\mathbb{R}^{n+1}, \langle ., . \rangle_p)$ which is $SO(p', n-p')$-equivariant with $p' = p$ or $p - 1$. Then $S$ is up to congruence (i.e. rotation and translation) one of the following:*

(i)  *either an affine hyperplane,*

(ii)  *or an open subset of a generalized catenoid*

$$\left\{ (rx, z) \in \mathbb{R}^{n+1} \mid x \in \mathbb{Q}_{p',1}^{n-1}, (r, z) \in \Gamma \right\},$$

*where $\Gamma := \gamma(\mathbb{R})$ is the image of a complete, embedded curve $\gamma$ of the half-plane $P = (0, \infty) \times \mathbb{R}$ endowed with the metric $dr^2 + \epsilon dz^2$. The curve $\gamma$, which has curvature $\kappa$ with respect to the unit normal vector $\nu = (\nu_1, \nu_2)$, is solution of the equation*

$$\kappa - (n-1)\frac{\nu_1}{r} = 0. \tag{3.3}$$

**Remark 15.**

(i)  In the Riemannian case, only the case $(\epsilon, \epsilon') = (1, 1)$ occurs and there is only one generalized catenoid.

(ii) In the Minkowski case, we have to look at both cases $\langle.,.\rangle_1$ and $\langle.,.\rangle_n$. There are however five, not six solutions: the case $p = p' = n$ is excluded, since the quadric $\mathbb{Q}_{n,1}^n$ is the empty set. There are two hypersurfaces with definite metric ("spacelike") when $(p', \epsilon, \epsilon') = (0, -1, 1)$ and $(p', \epsilon, \epsilon') = (n-1, -1, -1)$ and the three other hypersurfaces have Lorentzian metric (they are usually called "timelike"). If $n = 2$, these five solutions are the five catenoids described in Chapter 2.

(iii) From dimension $n = 3$ and beyond, the metric may be neither definite nor Lorentzian, in which case there are exactly six different generalized catenoids.

## 3.3 Equivariant minimal hypersurfaces in pseudo-space forms

### 3.3.1 *Totally umbilic hypersurfaces in pseudo-space forms*

We consider again the space form $\mathbb{Q}_{p,1}^{n+1}$, which from now on will be the ambient space. In the remainder of the chapter, we shall denote it by $\mathbb{Q}$ for brevity, assuming unlikely a confusion with the set of rational numbers. In order to simplify further notation, we consider $\mathbb{R}^{n+2} = \mathbb{R}^{n+1} \oplus \mathbb{R}$ with the metric

$$\langle.,.\rangle_p := \langle.,.\rangle_{p'} + \epsilon dx_{n+2}^2.$$

The space $\mathbb{Q}$ is endowed with the induced metric $\langle.,.\rangle_p$ and we shall denote by $D$ (instead of $\nabla$ in the previous section) the corresponding Levi-Civita connection. Since $SO(p, n + 2 - p)$ acts transitively on $\mathbb{Q}$, its only orbit is the whole space $\mathbb{Q}$. It is then natural to look at the action of

$$\widetilde{SO}(p', n + 1 - p') := \left\{ \begin{pmatrix} M & 0 \\ 0 & 1 \end{pmatrix} \Big| M \in SO(p', n + 1 - p') \right\},$$

where $p' = p - (1 - \epsilon)/2$. Observe that this subgroup fixes the "poles" $(0, ..., 0, \pm 1)$ of $\mathbb{Q}$. As we have seen in Section 3.2, the orbits of $\widetilde{SO}(p', n+1-p')$ in $\mathbb{R}^{n+2}$ take the form $(\mathbb{Q}_{p',c}^n, z)$. In particular, the set $\mathcal{Q}^z := (\mathbb{Q}_{p',c}^n, z)$, where $c := 1 - \epsilon z^2$, is contained in $\mathbb{Q}$. It is actually the intersection of $\mathbb{Q}$ with the affine hyperplane

$$\{\bar{x} \in \mathbb{R}^{n+2} | \langle \bar{x}, (0, ..., 0, 1) \rangle_p = z\}.$$

In the positive case $p = p' = 0, \epsilon = 1$, the sets $\mathcal{Q}^z$ are empty if $z^2 > 1$ and reduced to a point if $z = \pm 1$. In all other cases $\mathcal{Q}^z$ are smooth hypersurfaces of $\mathbb{Q}$. Moreover, in the indefinite case $p' \neq 0, \epsilon = 1$, the induced metric of

$Q^z$ is degenerate if $z = \pm 1$. We shall exclude all these cases in the following. Finally, we set $\epsilon_c := \frac{c}{|c|} = \pm 1$. Observe that the squared norm of the two unit normal vectors to $Q^z$ in $Q$ is $\eta := \epsilon \epsilon_c$.

The hyperplane just mentioned before is orthogonal to a non-null direction. It is natural, if the metric is indefinite, to consider also the case of the intersection of $Q$ with a hyperplane orthogonal to a null direction. The metric must be indefinite in order to enjoy null directions, and there is no loss of generality in assuming that it takes the form $\langle ., . \rangle_p := \langle ., . \rangle_{p-1} - dx_{n+1}^2 + dx_{n+2}^2$, according to the decomposition $\mathbb{R}^{n+2} = \mathbb{R}^n \oplus \mathbb{R}^2$. We then introduce, for $z \neq 0$, the *horosphere*:

$$\mathcal{H}^z := Q \cap \{\bar{x} \in \mathbb{R}^{n+2} | \langle \bar{x}, (0, ..., 0, 1, 1) \rangle_p = z\}.$$

The next theorem shows that the totally umbilic hypersurfaces of $Q$ are essentially the quadrics $Q^z$ and the horospheres $\mathcal{H}^z$:

**Theorem 14.** *The quadric $Q^z$ is a totally umbilic hypersurface of $Q$ and has constant curvature $\kappa = \dfrac{z}{\sqrt{|c|}}$ with respect to the unit normal vector field $\bar{N}(\bar{x}) := (-\dfrac{z}{\sqrt{|c|}}x, \eta\sqrt{|c|})$. In particular $Q^0$ is totally geodesic. Analogously the horosphere $\mathcal{H}^z$ is totally umbilic and has constant curvature $\kappa = 1$ with respect to the unit normal field $\bar{N}(\bar{x}) := -\bar{x} + \frac{1}{z}(0, ..., 0, 1, 1)$. Conversely, if $S$ is a connected, totally umbilic hypersurface of $Q$, its mean curvature is constant and $S$ is contained is an open subset of the intersection of $Q$ with an affine hyperplane (a linear hyperplane if $S$ is totally geodesic). In particular $S$ is, up to rotation, an open subset of one of the examples above.*

**Remark 16.** This result holds true in dimension $n = 1$ and therefore provides a classification of curves with constant curvature in the surface forms $Q_{p,1}^2$. One-dimensional horospheres are called *horocycles*.

*Proof of Theorem 14.* For sake of brevity, set, as in Section 3.2, $r := |c|^{1/2}$. In particular $c := \epsilon_c r^2$. A curve of $Q^z$ takes the form $\bar{x}(s) = (x(s), z)$, where $x(s)$ is a curve of $Q_{p',c}^n$. Hence the tangent vector to a point $\bar{x} := (x, z)$ of $S$, where $x \in Q_{p',c}^n$, takes the form $\bar{X} = (X, 0)$, where $X \in T_x Q_{p',c}^n$ and we have $T_{\bar{x}}Q^z = (T_x Q_{p',c}^n, 0)$. Recalling from the proof of Theorem 12 that $N := -|c|^{-1/2}x = -\frac{1}{r}x$ is a unit normal vector to $Q_{p',c}^n$ (in $\mathbb{R}^{n+1}$!), it follows that a unit normal vector to $Q^z$ (in $Q$!) must take the form $\bar{N} = (aN, b)$, with real numbers $a$ and $b$ satisfying

$$\begin{cases} |\bar{N}|_p^2 = \epsilon_c a^2 + \epsilon b^2 = \eta \\ \langle \bar{N}, \bar{x} \rangle_p = -\epsilon_c ar + \epsilon bz = 0. \end{cases}$$

A suitable choice of unit normal vector is therefore $\bar{N} = (zN, \eta r) = (-\frac{z}{r}x, \eta r)$. It remains to calculate the corresponding shape operator, using again Theorem 12:

$$D_{\bar{X}}\bar{N} = (D_X zN, D_X(\eta r)) = -\frac{z}{r}(D_X x, 0) = -\frac{z}{r}(X, 0) = -\kappa \bar{X},$$

hence $A_{\bar{N}}\bar{X} = \kappa \bar{X}$, i.e. $\mathcal{Q}^z$ is totally umbilic, and in particular totally geodesic for vanishing $z$.

We now proceed to calculate the shape operator of the horospheres. Given a point $\bar{x} = (x, x_{n+1}, x_{n+2})$ of $\mathcal{H}^z \subset \mathcal{Q} \subset \mathbb{R}^{n+2}$ with $x \in (\mathbb{R}^{n-1}, \langle \cdot, \cdot \rangle_{p-1})$, using that $-x_{n+1} + x_{n+2} = z$ and

$$1 = |x|_{p-1}^2 - x_{n+1}^2 + x_{n+1}^2 = |x|_{p-1}^2 + z(x_{n+1} + x_{n+2}),$$

we obtain

$$\begin{cases} x_{n+1} = \frac{-z+z^{-1}}{2} - r(x) \\ x_{n+2} = \frac{z+z^{-1}}{2} - r(x), \end{cases}$$

where we set $r(x) := \frac{|x|_{p-1}^2}{2z}$. It follows that a parametrization of $\mathcal{H}^z$ is

$$\begin{aligned} f : I \times \mathbb{R}^{n-1} &\to \mathcal{Q} \\ (s, x) &\mapsto (x, \frac{-z+z^{-1}}{2} - r(x), \frac{z+z^{-1}}{2} - r(x)). \end{aligned}$$

If $X \in T_x \mathbb{R}^{n-1} = \mathbb{R}^{n-1}$, then $\bar{X} := \left( X, -\frac{\langle x, X \rangle_{p-1}}{z}, -\frac{\langle x, X \rangle_{p-1}}{z} \right)$ is tangent to $\mathcal{H}^z$. Next an easy calculation proves that

$$\bar{N} := -\bar{x} + (0, \frac{1}{z}, \frac{1}{z}) = \left( -x, \frac{z+z^{-1}}{2} + r(x), \frac{-z+z^{-1}}{2} + r(x) \right)$$

is a unit normal vector along $\mathcal{H}^z$ and that

$$D_{\bar{X}}\bar{N} = \left( -X, \frac{\langle x, X \rangle_{p-1}}{z}, \frac{\langle x, X \rangle_{p-1}}{z} \right) = -\bar{X},$$

hence $\mathcal{H}^z$ is totally umbilic with constant curvature $\kappa = 1$.

Conversely, let $S$ be a totally umbilic hypersurface of $\mathcal{Q}$ and denote by $x$ a point of $S \subset \mathcal{Q} \subset \mathbb{R}^{n+2}$. Let $N$ be one unit normal vector of $S$ at $x$ in $\mathcal{Q}$. The hypersurface $S$ is totally umbilic if and only if there exists some function $\kappa(x)$ on $S$ such that $\nabla_X N = -\kappa X$, where $\nabla$ denotes the Levi-Civita connection on $S$ and $X$ is any tangent vector field. We first show that, as in Theorem 12, $\kappa$ is constant. Differentiating the identities $D_X N = -\kappa(x)X$ and $D_Y N = -\kappa(x)Y$ with respect to $Y$ and $X$ respectively, we have

$$D_Y D_X N = -Y(\kappa)X - \kappa D_Y X$$

and

$$D_X D_Y N = -X(\kappa)Y - \kappa D_X Y.$$

Subtracting these two equations, we get

$$D_Y D_X N - D_X D_Y N = X(\kappa)Y - Y(\kappa)X + \kappa[X, Y].$$

Since $D_{[X,Y]}N = -\kappa[X, Y]$, it follows that:

$$D_Y D_X N - D_X D_Y N - D_{[X,Y]}N = X(\kappa)Y - Y(\kappa)X.$$

The left hand side term of this expression is nothing but the curvature tensor $R^D$ evaluated at the triple $(X, Y, N)$. It follows from Equation (3.1) of Proposition 7, that

$$D_Y D_X N - D_X D_Y N - D_{[X,Y]}N = \langle X, N \rangle_p Y - \langle Y, N \rangle_p X = 0.$$

Hence by choosing $X$ and $Y$ not collinear, we get that both $X(\kappa)$ and $Y(\kappa)$ vanish. Since $S$ is assumed to be connected, $\kappa$ is constant.

We now introduce the vector field $V := \kappa x + N$ along $S$. First observe that since $S$ is totally umbilic, we have

$$D_X V = \kappa X + D_X N = 0,$$

hence $V$ is constant on $S$. Moreover, we have

$$\langle V, x \rangle_p = \kappa |x|_p^2 + \langle N, x \rangle_p = \kappa.$$

Therefore $S$ is contained in the set $\{x \in \mathbb{R}^{n+2} | \langle V, x \rangle_p = \kappa\}$ which is a linear hyperplane of $\mathbb{R}^{n+2}$ if $\kappa$ vanishes, and an affine hyperplane otherwise. If $V$ is non-null, we may assume, after a suitable change of coordinates, that it takes the form $V = (0, ..., 0, \lambda)$, where $\lambda$ is a non-vanishing real number. Hence

$$\{x \in \mathbb{R}^{n+2} | \langle V, x \rangle_p = \kappa\} = \mathcal{Q}^z \quad \text{with} \quad z := \frac{\kappa}{\lambda}.$$

Suppose now that $V$ is a null vector. Then $|V|_p^2 = \kappa^2 |x|_p^2 + |N|_p^2 = 0$, hence $\kappa^2 = -|N|_p^2$ and therefore $\kappa = 1$ or $-1$. We may assume that $\kappa = 1$, replacing $N$ by $-N$ if necessary. Moreover, after a change of coordinates, we may assume that $V$ takes the form $V = (0, ..., 0, \lambda, \lambda)$, where $\lambda$ is a non-vanishing real number. Hence

$$\{x \in \mathbb{R}^{n+2} | \langle V, x \rangle_p = 1\} = \mathcal{H}^z \quad \text{with} \quad z := \frac{1}{\lambda}. \qquad \square$$

### 3.3.2  *Equivariant hypersurfaces in pseudo-space forms*

We have described in the previous sub-section the hypersurfaces of $\mathbb{Q}$ which are $SO(p, n+1-p)$-equivariant. In order to get a construction equivalent to that done in pseudo-Euclidean space, which resulted in the description of generalized catenoids, we consider smaller subgroups, whose non-degenerate orbits are $(n-1)$-dimensional submanifolds, and therefore, whose equivariant hypersurfaces will be described by a one-parameter family of such orbits. Set

$$\widetilde{SO}(p', n - p') := \left\{ \begin{pmatrix} M & 0 & 0 \\ 0 & 1 & 0 \\ 0 & 0 & 1 \end{pmatrix} \,\middle|\, M \in SO(p', n - p') \right\},$$

where $p' \in \{p, p-1, p-2\}$ and the metric on $\mathbb{R}^{n+2} = \mathbb{R}^n \oplus \mathbb{R}^2$ takes the form

$$\langle \cdot, \cdot \rangle_p := \langle \cdot, \cdot \rangle_{p'} + \epsilon' dx_{n+1}^2 + \epsilon'' dx_{n+2}^2,$$

with

(i) $p' = p$ if $(\epsilon', \epsilon'') = (1, 1)$;
(ii) $p' = p - 1$ if $(\epsilon', \epsilon'') = (-1, 1)$ or $(1, -1)$;
(iii) $p' = p - 2$ if $(\epsilon', \epsilon'') = (-1, -1)$.

An $SO(p', n-p')$-equivariant hypersurface of $\mathbb{Q}$ may be locally parametrized by an immersion of the form

$$
\begin{aligned}
f : I \times \mathbb{Q}_{p',\epsilon}^{n-1} &\to \mathbb{Q} \\
(s, x) &\mapsto (r(s)x, z_1(s), z_2(s)),
\end{aligned}
$$

where

$$\epsilon r^2 + \epsilon' z_1^2 + \epsilon'' z_2^2 = 1, \tag{3.4}$$

where $r(s), z_1(s)$ and $z_2(s)$ are three real-valued functions defined on the interval $I$, and $r$ is assumed to be strictly positive. In other words, $\gamma(s) = (r(s), z_1(s), z_2(s))$ is a parametrized curve in the "right hemisphere" $\Sigma := \{r > 0\} \cap \mathbb{Q}_{p'',1}^2$ of the surface form $\mathbb{Q}_{p'',1}^2 \subset (\mathbb{R}^3, \langle \cdot, \cdot \rangle_{p''})$, where $p'' = p - p' - \frac{\epsilon+1}{2}$.

**Remark 17.** Unlike the case of pseudo-Euclidean space, we cannot assume that $\epsilon = 1$, since replacing the metric $\langle \cdot, \cdot \rangle_p$ by $\langle \cdot, \cdot \rangle_{n-p}$ would result in an immersion valued in $\mathbb{Q}_{p,-1}^n$ instead of $\mathbb{Q}$.

We introduce coordinates $(s_1, ..., s_{n-1})$ in $\mathbb{Q}^{n-1}_{p',\epsilon}$, so that $(s_0 := s, s_1, ..., s_{n-1})$ are local coordinates on $I \times \mathbb{Q}^{n-1}_{p',\epsilon}$. We set $X_i := d\iota(\partial_{s_i})$, where $\iota$ is the canonical immersion of $\mathbb{Q}^{n-1}_{p',\epsilon}$ in $\mathbb{R}^n$. In particular $(X_1, ..., X_{n-1})$ is a frame of tangent vectors along $\mathbb{Q}^{n-1}_{p',\epsilon}$.

Computing

$$f_s = (r'x, z'_1, z'_2) \quad \text{and} \quad df(\partial_{s_i}) = (rX_i, 0, 0),$$

it is easy to check that $f$ is an immersion if and only if $\gamma'$ do not vanish (and $r > 0$, which we already assumed). In particular, the curve $\gamma$ is regular. We claim that the geometry of the immersion $f$ is described by that of $\gamma$ in $(\Sigma, \langle ., .\rangle_{p''})$: denoting by $g$ the metric on $\mathbb{Q}^{n-1}_{p',\epsilon}$ the coefficients of the induced metric $\bar{g}$ on $I \times \mathbb{Q}^{n-1}_{p',\epsilon}$ are:

$$\bar{g}_{00} = |\gamma'|^2_{p''}, \quad \bar{g}_{0i} = 0 \quad \text{and} \quad \bar{g}_{ij} = r^2 g_{ij}.$$

In other words,

$$\bar{g} = |\gamma'|^2_{p''} ds^2 + r^2 g.$$

It follows that the immersion is degenerate if and only if $\bar{g}_{00}$ vanishes, i.e. $\gamma$ is degenerate (null). Hence we assume from now on that $\gamma$ is non-null and that $s$ is the arc length parameter of the curve

Moreover, denoting by $\nu = (\nu_1, \nu_2, \nu_3)$ a unit normal vector of $\gamma$ in $\Sigma$, the Frénet equations for the curve $\gamma$ are (see Chapter 1, Section 1.2.3):

$$\begin{cases} \gamma'' = \epsilon_0 \kappa \nu \\ \nu' = -\kappa \gamma', \end{cases}$$

where $\epsilon_0 = -1$ if $p'' = 1$ and $\epsilon_0 = 1$ otherwise (we moreover have $\epsilon_0 = |\nu|^2_{p''} |\gamma'|^2_{p''} = \epsilon \epsilon' \epsilon''$).

We now proceed to compute the second fundamental form of $f$. Observe that a unit normal vector to $f$ is given by

$$N = (\nu_1 x, \nu_2, \nu_3).$$

Next we compute, using the Frénet equations:

$$D_{\partial_s} N = (\nu'_1 x, \nu'_2, \nu'_3) = -\kappa(r'x, z'_1, z'_2) = -\kappa f_s,$$
$$D_{\partial_{s_i}} N = (\nu_1 X_i, 0, 0).$$

Hence the coefficients of the second fundamental form $\bar{h}$ of $f$ are

$$\bar{h}_{00} = -\bar{g}(D_{\partial_s} N, \partial_s) = \kappa |f_s|^2_p = \kappa \bar{g}_{00},$$
$$\bar{h}_{0i} = -\bar{g}(D_{\partial_{s_i}} N, \partial_s) = -\langle \nu_1 X_i, r'x \rangle_{p'} = 0,$$
$$\bar{h}_{ij} = -\bar{g}(D_{\partial_{s_i}} N, \partial_{s_j}) = -\langle \nu_1 X_i, r X_j \rangle_{p'} = -\nu_1 r \bar{g}_{ij}.$$

In particular, the shape operator is diagonalisable and its principal curvatures are $\kappa_0 = \kappa$ with multiplicity one and $\kappa_1 = -\frac{\nu_1}{r}$ with multiplicity $n-1$, and the mean curvature is

$$H = \frac{1}{n}\left(\kappa - (n-1)\frac{\nu_1}{r}\right). \tag{3.5}$$

Hence in order to find equivariant minimal immersions, we have to solve

$$\kappa - (n-1)\frac{\nu_1}{r} = 0. \tag{3.6}$$

Since the equation is formally identical to that of Euclidean case, i.e. Equation (3.3), the function $E(\gamma, \gamma') = \nu_1 r^{n-1}$ is still a first integral of Equation (3.6).

### 3.3.3  *Totally geodesic and isoparametric solutions*

We now discuss the existence of special solutions. The simplest solution arises assuming that $\kappa$ (hence $\nu_1$ as well) vanishes. In particular the corresponding hypersurface is totally geodesic. From Theorem 14 we know that a geodesic of the $\mathbb{Q}^2_{p'',1}$ is the intersection of the surface with a linear plane $\nu^\perp$, the unit normal vector $\nu$ of the geodesic being constant. Since here $\nu_1$ vanishes, this plane $\nu^\perp$ is vertical, so up to rotation in the plane $(z_1, z_2)$, we may assume that $\gamma$ is one of the geodesics $\{z_1 = 0\}$ or $\{z_2 = 0\}$ of $\mathbb{Q}^2_{p'',1}$. The corresponding totally geodesic hypersurface is the intersection of $\mathbb{Q}$ with one the hyperplanes $\{x_{n+1} = 0\}$ or $\{x_{n+2} = 0\}$ of $\mathbb{R}^{n+2}$.

We get other, more interesting special solutions assuming that $r$ is constant (hence $\nu_1$ as well). In this case $\gamma$ is contained in the intersection of $\mathbb{Q}^2_{p'',1}$ with an affine plane $\{r = \bar{r}\}$, where $\bar{r}$ is some constant to determine. For further use we set

$$c := \epsilon' z_1^2 + \epsilon'' z_2^2 = 1 - \epsilon^2 \bar{r}^2.$$

Hence by Theorem 14 (see also Remark 16), $\gamma$ has constant curvature $\kappa = \frac{\bar{r}}{\sqrt{|c|}}$ and $\nu_1 = \eta\sqrt{|c|}$, where $\eta := |\nu|^2_{p''}$ (beware to the notations that might be confusing: the quantity $\bar{r}$ corresponds to the quantity $z$ of Theorem 14, the first coordinate here corresponding to the last one of the theorem). In particular, the two principal curvatures of $f$ are constant, and the corresponding hypersurface is isoparametric (see Exercise 2 of Chapter 1). Hence Equation (3.6) amounts to

$$\frac{\bar{r}}{\sqrt{|1 - \epsilon\bar{r}^2|}} - (n-1)\frac{\eta\sqrt{|1 - \epsilon\bar{r}^2|}}{\bar{r}} = 0,$$

so

$$\bar{r}^2 = \eta(n-1)|1 - \epsilon\bar{r}^2|. \tag{3.7}$$

In particular, we must have $\eta = 1$ i.e. the unit normal vector to $\gamma$ is positive, which excludes the case of the hyperbolic plane $p'' = 2$ and leaves us with the cases of the sphere and the de Sitter surface. Moreover, the case $\epsilon = -1$ is excluded as well, since the equation $\bar{r}^2 = (n-1)|1 + \bar{r}^2|$ has no real solution.

**First case:** the sphere ($p'' = 0$).

Here, $c = 1 - \bar{r}^2 > 0$ so $\bar{r} = \sqrt{\frac{n-1}{n}}$. Hence the constant curvature curve $\gamma$ solution of Equation (3.6) may be parametrized by

$$\gamma(t) = \left( \sqrt{\frac{n-1}{n}}, \sqrt{\frac{1}{n}}(\cos t, \sin t) \right) \in \mathbb{S}^2.$$

**Second case:** the de Sitter surface ($p'' = 1$).

If $c = 1 - \bar{r}^2 > 0$, then $\bar{r} = \sqrt{\frac{n-1}{n}}$ again and we get

$$\gamma(t) = \left( \sqrt{\frac{n-1}{n}}, \sqrt{\frac{1}{n}}(\sinh t, \cosh t) \right) \in d\mathbb{S}^2.$$

If $1 - \bar{r}^2 < 0$, then $\bar{r} = \sqrt{\frac{n-1}{n-2}}$ and we get

$$\gamma(t) = \left( \sqrt{\frac{n-1}{n-2}}, \frac{1}{\sqrt{n-2}}(\cosh t, \sinh t) \right)$$

but this case is excluded since it implies $\eta = -1$.

In order to discuss the general solution of Equation (3.6), it is necessary to deal separately with the different values of the triple $(\epsilon, \epsilon', \epsilon'')$. This will be the subject of the next sections.

### 3.3.4   *The spherical case* $(\epsilon, \epsilon', \epsilon'') = (1, 1, 1)$

Here $\Sigma$ is the right hemisphere $\mathbb{S}^2 \cap \{r > 0\}$ (in particular $p' = p$ and $p'' = 0$). We introduce spherical coordinates on $\Sigma$:

$$\gamma = (r, z_1, z_2) = (\cos\psi, \sin\psi \cos\varphi, \sin\psi \sin\varphi),$$

where $\psi, \varphi \in \mathbb{R}/2\pi\mathbb{Z}$, and the orthonormal tangent frame $(e_1, e_2)$ defined by

$$e_1 := (0, -\sin\varphi, \cos\varphi) \quad \text{and} \quad e_2 := (-\sin\psi, \cos\psi\cos\varphi, \cos\psi\sin\varphi),$$

and we define $\theta$ to be the angle of the tangent $\gamma'$ with respect to the direction $e_1$, i.e. $\theta$ is the $\mathbb{R}/2\pi\mathbb{Z}$-valued function determined by $\gamma' = \cos\theta\, e_1 + \sin\theta\, e_2$. Since $\gamma' = \varphi' \sin\psi e_1 + \psi' e_2$, it follows that

$$\varphi' \sin\psi = \cos\theta \quad \text{and} \quad \psi' = \sin\theta.$$

Moreover, choosing as unit normal vector $\nu := \sin\theta\, e_1 - \cos\theta\, e_2$, we get $\nu_1 = \sin\psi\cos\theta$, so finally

$$E = \cos\theta \sin\psi \cos^{n-1}\psi.$$

From this expression we can easily draw the projections of the solution $(\gamma, \gamma')$ onto the plane $(\theta, \psi)$, which are of three types:

(i) the vertical segment $\psi = 0[\pi]$. The corresponding solution $\gamma$ has energy level $E = 0$ and is a geodesic; this is the totally geodesic case;

(ii) the point $(0[\pi], \psi_0,)$, where $\psi_0 := \arccos(\sqrt{\frac{n-1}{n}})$. Its energy level is $E_0 := e(\psi_0)$. The corresponding $\gamma$ has constant curvature; this is the isoparametric case discussed in the previous section;

(iii) a closed curve winding around the fixed point $(0[\pi], \psi_0)$ and contained in the open subset

$$\{-\pi/2 + k\pi < \theta < \pi/2 + k\pi, 0 < \psi < \pi/2\},$$

where $k \in \mathbb{Z}$. Since the equation is periodic, we may assume without loss of generality that $k = 0$.

In order to understand the global properties of the curves $\gamma$ corresponding to solutions of type (iii), we need to evaluate the variable $\varphi$, which is known implicitly via the equation $\varphi' \sin\psi = \cos\theta$. In particular we set $\Phi(E) := \int_s^{s+T} \varphi'(\sigma)d\sigma$, where $T$ is the period of $(\theta, \psi)$ and we have the following result:

**Lemma 6.** *The curve $\gamma$ corresponding to the solution of type (iii) with energy level $E$ is closed if and only if $\Phi(E)$ takes the form $\Phi(E) = \frac{a}{b}2\pi$, where $\frac{a}{b} \in \mathbb{Q}$, and moreover has self-intersections if $a \neq 1$.*

*Proof.* Since the domain of parametrization of $\gamma$ is $\mathbb{R}$, its image is a closed curve of $\mathbb{S}^2 \cap \{r > 0\}$ if and only if $\gamma$ is periodic. On the one hand $\psi$ is periodic of period $T$, hence $(\cos\psi, \sin\psi)$ as well, and on the other one, if $\Phi(E) = \frac{a}{b}2\pi$, we have $\varphi(s + bT) = \varphi(s) + 2a\pi$, so $(\cos\varphi, \sin\varphi)$ is periodic of period $bT$. Therefore if $\Phi(E)$ is rationally related to $2\pi$, the curve $\gamma$ is periodic and its image is a closed curve. Conversely, if $\Phi(E) \notin 2\pi\mathbb{Q}$, the curve $\gamma(s)$ has no period and the set $\gamma(\mathbb{R})$ is a dense subset of $\mathbb{S}^2 \cap \{r > 0\}$ whose closure is the annulus

$$\left\{ (\cos\psi, \sin\psi\cos\varphi, \sin\psi\cos\varphi) \Big| \, \psi \in \left( \inf_{s\in\mathbb{R}} \psi(s), \sup_{s\in\mathbb{R}} \psi(s) \right) \right\}.$$

We now assume that $\Phi(E) = \frac{a}{b}2\pi$, so that $\gamma$ has period $bT$, and we discuss the embeddedness of $\gamma$. Observe first that the function $\varphi$ is monotone. Moreover, if $a = 1$, the range of $\varphi$ over a period of $\gamma$ is exactly $2\pi$, so the curve makes a wind around the "north pole" $(1, 0, 0)$ and the image of an interval $[s, s + bT]$ by $\gamma$ is a simple curve. If $a \neq 1$, the curve $\gamma$ makes $a$ winds around the north pole so intuitively $\gamma$ must cross itself to close up. To give a rigorous proof of this claim, we set $\sigma(t)$ to be the function inverse of $\varphi$, i.e. such that $\varphi(\sigma(t)) = t$ and apply the intermediate value theorem to the function $\xi(t) := \psi \circ \sigma(t) - \psi \circ \sigma(t + 2\pi)$ on an interval $(t_1, t_2)$ of $\mathbb{R}$ such that $\psi \circ \sigma(t_1) = \inf_{s\in\mathbb{R}} \psi(s)$ and $\psi \circ \sigma(t_2) = \sup_{s\in\mathbb{R}} \psi(s)$. Since $a \neq 1$, we have $\sigma(t + 2\pi) - \sigma(t) \notin bT\mathbb{Z}$, hence $\xi(t_1) < 0$ and $\xi(t_2) > 0$. Thus there exists $t_0 \in (t_1, t_2)$ such that $\xi(t_0) = 0$. Therefore, $\psi \circ \sigma(t_0) = \psi \circ \sigma(t_0 + 2\pi)$, so the curve $\gamma$ has a point of self-intersection at $\gamma(\sigma(t_0)) = \gamma(\sigma(t_0 + 2\pi))$.$\square$

It turns out to be difficult to evaluate the quantity $\Phi(E)$, but it can be proved that it ranges an open interval, hence there are countably many periods rationally related to $2\pi$. On the other hand, none of them takes the form $\Phi(E) = \frac{1}{b}2\pi$:

**Proposition 8.** *The variation of $\varphi$ on a period of $\psi$, i.e. the number $\Phi(E)$, ranges the open interval $(\pi, \sqrt{2}\pi)$ and belongs to the open interval $(\pi, \sqrt{\frac{n-1}{n}}2\pi)$. Hence there is a countable family of closed solutions of Equation (3.6), all with self-intersections. The other solutions are non properly immersed curves.*

The proof requires hard computations that are omitted. The facts that $\lim_{E\to E_0} \Phi(E) = \sqrt{2}\pi$, $\lim_{E\to 0} \Phi(E) = \pi$ and $\Phi(E) > \pi$ have been proved in [Otsuki (1970)], while [Furuya (1971)] showed the inequality $\Phi(E) \leq \sqrt{\frac{n-1}{n}}2\pi$ (see also [Brito, Leite (1990)]). It is still conjectured that $\Phi(E) < \sqrt{2}\pi$.

### 3.3.5    The "elliptic hyperbolic" case $(\epsilon, \epsilon', \epsilon'') = (1, -1, -1)$

Here the surface $\Sigma$ is the following model of the hyperbolic plane $\{(r, z_1, z_2)|r^2 - z_1^2 - z_2^2 = 1, r > 0\}$ (in particular $p' = p - 2$ and $p'' = 2$).

We introduce "elliptic coordinates" on $\Sigma$:

$$\gamma = (r, z_1, z_2) = (\cosh\psi, \sinh\psi\cos\varphi, \sinh\psi\sin\varphi),$$

where $\psi \in (0, \infty), \varphi \in \mathbb{R}/2\pi\mathbb{Z}$, and the orthonormal tangent frame $(e_1, e_2)$ along $(\mathbb{H}^2, \langle.,.\rangle_2)$ defined by

$$e_1 := (0, -\sin\varphi, \cos\varphi) \quad \text{and} \quad e_2 := (\sinh\psi, \cosh\psi\cos\varphi, \cosh\psi\sin\varphi).$$

Next we define $\theta$ to be the angle of the velocity vector $\gamma'$ with respect to $(e_1, e_2)$, i.e. the $\mathbb{R}/2\pi\mathbb{Z}$-valued function $\theta$ determined by $\gamma' = \cos\theta\, e_1 + \sin\theta\, e_2$. It follows that

$$\varphi' \sinh\psi = \cos\theta \quad \text{and} \quad \psi' = \sin\theta.$$

Moreover, choosing as unit normal vector

$$\nu := -\sin\theta\, e_1 + \cos\theta\, e_2,$$

we have $\nu_1 = \cos\theta \sinh\psi$. Hence

$$E = \cos\theta \sinh\psi \cosh^{n-1}\psi.$$

In particular the coordinate $\psi$ has a minimum $\psi_0$ at $\theta = 0$ and the projections of the solutions $(\gamma, \gamma')$ on the plane $(\theta, \psi)$ have two unbounded ends $(\theta, \psi) \to (\pm\pi/2, \infty)$. Without loss of generality, we may choose the parameter $s$ in such a way that $\psi_0 = \psi(0)$, and assume as well that $\varphi(0) = 0$. It follows that $\psi(-s) = \psi(s)$. Therefore, the curve $\gamma$ has a self-intersection if and only if there exists $s_0 \in \mathbb{R}$ such that $\varphi(-s) - \varphi(s) \in 2\pi\mathbb{Z}$. The following lemma proves that this cannot happen and therefore that $\gamma$ is embedded:

**Lemma 7.**

$$\Phi(E) := \lim_{s \to \infty} \varphi(s) - \lim_{s \to -\infty} \varphi(s) < \pi.$$

*Proof.* From the fact that

$$0 = \frac{dE}{ds} = -\theta' \sin\theta \sinh\psi \cosh^{n-1}\psi + \psi' \cos\theta \frac{d}{d\psi}\left(\sinh\psi \cosh^{n-1}\psi\right) = 0,$$

we deduce that

$$\theta' = \cos\theta\left(\cosh\psi + (n-1)\sinh\psi\tanh\psi\right).$$

On the other hand,

$$\varphi(s) - \varphi(s') = \int_{s'}^{s} \varphi'(\sigma)d\sigma = \int_0^s \frac{\cos\theta}{\sinh\psi}d\sigma,$$

so we conclude, using the fact that $\cosh\psi + (n-1)\sinh\psi\tanh\psi > 1$,

$$\lim_{s\to\infty}\varphi(s) - \lim_{s\to-\infty}\varphi(s) = \int_{-\infty}^{\infty}\frac{\cos\theta}{\sinh\psi}\,d\sigma$$

$$= \int_{-\pi/2}^{\pi/2}\frac{\varphi'}{\theta'}\,d\theta$$

$$= \int_{-\pi/2}^{\pi/2}\frac{d\theta}{\cosh\psi + (n-1)\sinh\psi\tanh\psi}$$

$$< \pi \qquad\qquad \Box$$

### 3.3.6   The "hyperbolic hyperbolic" case $(\epsilon, \epsilon', \epsilon'') = (-1, -1, 1)$

Here the surface $\Sigma$ is the right "hemisphere" of the following model of the hyperbolic plane $\{(r, z_1, z_2)| - r^2 - z_1^2 + z_2^2 = 1, r > 0\}$; the rôles of $z_1$ and $z_2$ being equivalent in the equation, the case $(\epsilon, \epsilon', \epsilon'') = (-1, 1, -1)$ is similar. We introduce "hyperbolic coordinates" on $\Sigma$:

$$\gamma = (r, z_1, z_2) = (\sinh\psi, \cosh\psi\sinh\varphi, \cosh\psi\cosh\varphi),$$

where $\psi \in (0, \infty)$ and $\varphi \in \mathbb{R}$, and the orthonormal tangent frame $(e_1, e_2)$ on $(\mathbb{H}^2, \langle.,.\rangle_2)$ defined by

$$e_1 := (0, \cosh\varphi, \sinh\varphi) \quad \text{and} \quad e_2 := (\cosh\psi, \sinh\psi\sinh\varphi, \sinh\psi\cosh\varphi).$$

Next we define $\theta$ to be the angle of the tangent $\gamma'$ with respect to $(e_1, e_2)$, i.e. $\gamma' = \cos\theta\, e_1 + \sin\theta\, e_2$. It follows that

$$\varphi'\cosh\psi = \cos\theta \quad \text{and} \quad \psi' = \sin\theta.$$

Moreover, choosing as unit normal vector $\nu := -\sin\theta\, e_1 + \cos\theta\, e_2$, we have $\nu_1 = \cos\theta\coth\psi$, so finally

$$E = \cos\theta\cosh\psi\sinh^{n-1}\psi.$$

The profile of the projections of the solutions $(\gamma, \gamma')$ on the plane $(\theta, \psi)$ is the same than in the previous case. In particular the curves $\gamma$ have two unbounded ends. There is, however, no period problem here for the variable $\varphi$ since the coordinates are $(\sinh\psi, \cosh\psi\sinh\varphi, \cosh\psi\cosh\varphi)$: $\varphi$ is strictly increasing and thus the corresponding curves $\gamma$ are embedded.

### 3.3.7  The "elliptic" de Sitter case $(\epsilon, \epsilon', \epsilon'') = (-1, 1, 1)$

The surface is the right "hemisphere" of the de Sitter surface

$$\{(r, z_1, z_2)| -r^2 + z_1^2 + z_2^2 = 1, r > 0\}$$

(in particular $p' = p - 1$ and $p'' = 1$). We introduce "elliptic coordinates" on $\Sigma$:

$$\gamma = (r, z_1, z_2) = (\sinh\psi, \cosh\psi\cos\varphi, \cosh\psi\sin\varphi),$$

where $\psi \in (0, \infty), \varphi \in \mathbb{R}/2\pi\mathbb{Z}$, and the orthonormal tangent frame $(e_1, e_2)$ on $(d\mathbb{S}^2, \langle ., . \rangle_1)$, defined by

$$e_1 := (0, -\sin\varphi, \cos\varphi) \quad \text{and} \quad e_2 := (\cosh\psi, \sinh\psi\cos\varphi, \sinh\psi\sin\varphi).$$

Observe that $e_1$ is positive and $e_2$ is negative. Hence we define $\theta$ to be the hyperbolic angle of the tangent $\gamma'$ with respect to $(e_1, e_2)$, i.e. the $\mathbb{R}$-valued function $\theta$ determined by $\gamma' = \cosh\theta\, e_1 + \sinh\theta\, e_2$ or $\gamma' = \sinh\theta e_1 + \cosh\theta e_2$, depending of whether $\gamma'$ is positive or negative. It follows that

$$\varphi'\cosh\psi = \cosh\theta \quad \text{and} \quad \psi' = \sinh\theta$$

$$(\text{resp. } \varphi'\cosh\psi = \sinh\theta \quad \text{and} \quad \psi' = \cosh\theta).$$

Moreover, one choice of unit normal vector is $\nu = \sinh\theta\, e_1 + \cosh\theta\, e_2$ (or $\nu = \cosh\theta\, e_1 + \sinh\theta\, e_2$). In particular $\nu_1 = \sinh\theta\cosh\psi$ (resp. $\nu_1 = \cosh\theta\cosh\psi$, so finally

$$E = \cosh\theta\cosh\psi\sinh^{n-1}\psi$$

$$(\text{resp. } E = \sinh\theta\cosh\psi\sinh^{n-1}\psi).$$

#### 3.3.7.1  The positive case

In this case, the range of the coordinate $\psi$ is $(0, \psi_0)$, where $\psi_0$ is the only root of the equation $\cosh\psi\sinh^{n-1}\psi = E$. It follows that $\gamma$ is contained in the annulus $\{0 < r < \sinh\psi_0\}$ of the de Sitter surface, As $s$ tends to $\infty$, the curve approaches the equator $\{r = 0\}$ of $\Sigma$, while its velocity becomes asymptotically null and the arc length tends to infinity. As in Section 3.3.5, the embeddedness of $\gamma$ follows from an estimation of the total variation of the angle $\varphi$.

**Lemma 8.**

$$\Phi(E) := \left| \lim_{s\to\infty} \varphi(s) - \lim_{s\to-\infty} \varphi(s) \right| < \pi.$$

*Proof.* By a standard computation, one shows that $\Phi'(E) > 0$. On the other hand, we have $\lim_{E\to\infty} \Phi(E) = 2\int_0^\infty \frac{d\psi}{\cosh\phi} = \pi$, and the claims follows.  $\square$

Therefore, the curve $\gamma$ touches the equator at two end-points. At these two points the corresponding hypersurface becomes degenerate. This situation is similar to that of Section 3.2.5.

### 3.3.7.2    *The negative case*

Here, the coordinate $\psi$ is monotone, so the solutions $\gamma$ are embedded as well. The projections of the curves $(\gamma, \gamma')$ on the plane $(\theta, \psi)$ have two unbounded ends $(\theta, \psi) \to (\infty, 0)$ and $(\theta, \psi) \to (0, \infty)$. Moreover, the total variation of $\varphi$ is estimated as follows:

$$\lim_{s \to \infty} \varphi(s) - \lim_{s \to -\infty} \varphi(s) = \int_0^\infty \frac{\varphi'}{\psi'} d\psi$$

$$= \int_0^\infty \frac{d\psi}{\cosh \psi} \left( 1 + \frac{\cosh \psi \sinh^{n-1} \psi}{E^2} \right)^{-1/2}$$

$$\leq \int_0^\infty \frac{d\psi}{\cosh \psi}$$

$$\leq \pi/2.$$

In particular the curve $\gamma$ is embedded and contained in a "quadrant" $\{\varphi_0 \leq \varphi \leq \varphi_0 + \frac{\pi}{2}\}$ of the de Sitter surface. Moreover the curve $\gamma$ has a null endpoint at the equator $\{r = 0\}$ of the de Sitter surface, and an unbounded end asymptotic to a vertical geodesic. This situation is similar to that of Section 3.2.4.

### 3.3.8    *The "hyperbolic" de Sitter case $(\epsilon, \epsilon', \epsilon'') = (1, -1, 1)$*

Here the surface is the right "hemisphere" of the de Sitter surface $\{(r, z_1, z_2) | r^2 - z_1^2 + z_2^2 = 1, r > 0\}$ (in particular $p' = p - 1$ and $p'' = 1$). Since the rôles of $z_1$ and $z_2$ are equivalent in the equation, the case $(\epsilon, \epsilon', \epsilon'') = (1, 1, -1)$ is similar.

### 3.3.8.1    *The region $\{|z_1| > |z_2|\}$*

We introduce "hyperbolic coordinates" in the region $\{|z_1| > |z_2|\}$
$$\gamma = (r, z_1, z_2) = (\cosh \psi, \sinh \psi \cosh \varphi, \sinh \psi \sinh \varphi),$$
where $\psi, \varphi \in \mathbb{R}$, and the orthonormal frame $(e_1, e_2)$ on $(d\mathbb{S}^2, \langle ., . \rangle_1)$ defined by
$$e_1 := (0, \sinh \varphi, \cosh \varphi) \quad \text{and} \quad e_2 := (\sinh \psi, \cosh \psi \cosh \varphi, \cosh \psi \sinh \varphi).$$

As done before we consider the hyperbolic angle of $\gamma'$, i.e. the angle function $\theta$ defined by $\gamma' = \cosh \theta \, e_1 + \sinh \theta \, e_2$ or $\gamma' = \sinh \theta e_1 + \cosh \theta e_2$, depending on whether $\gamma'$ is positive or negative. It follows that

$$\varphi' = \frac{\cosh\theta}{\sinh\psi} \quad \text{and} \quad \psi' = \sinh\theta$$

$$\left(\text{resp. } \varphi' = \frac{\sinh\theta}{\sinh\psi} \quad \text{and} \quad \psi' = \cosh\theta\right).$$

Moreover, choosing as unit normal vector $\nu := \sinh\theta\, e_1 + \cosh\theta\, e_2$, (resp. $\nu := \cosh\theta\, e_1 + \sinh\theta\, e_2$, we obtain

$$E = \cosh\theta \sinh\psi \cosh^{n-1}\psi$$

$$(\text{resp. } E = \sinh\theta \sinh\psi \cosh^{n-1}\psi).$$

The profile of the projections of the curves solutions $(\gamma, \gamma')$ on the plane $(\theta, \psi)$ are similar (and even identical if $n = 2$) to the elliptic de Sitter case (Section 3.3.7), however here the coordinate $\varphi$ is a hyperbolic angle rather than a circular one. Moreover, analogous computations show that the total variation of $\varphi$ is infinite. Hence both in the positive and negative cases, the solutions $\gamma$ are embedded curves with two unbounded ends.

### 3.3.8.2 *The region* $\{|z_1| < |z_2|\}$

We introduce other coordinates in the region $\{|z_1| < |z_2|\}$

$$\gamma = (r, z_1, z_2) = (\cos\psi, \sin\psi \sinh\varphi, \sin\psi \cosh\varphi),$$

where $\psi \in (-\pi/2, \pi/2)$ and $\varphi \in \mathbb{R}$, and the orthonormal frame $(e_1, e_2)$ on $(d\mathbb{S}^2, \langle.,.\rangle_1)$ defined by

$$e_1 := (0, \cosh\varphi, \sinh\varphi) \quad \text{and} \quad e_2 := (-\sin\psi, \cos\psi \sinh\varphi, \cos\psi \cosh\varphi).$$

The hyperbolic angle $\theta$ of $\gamma'$ is defined by $\gamma' = \sinh\theta\, e_1 + \cosh\theta\, e_2$ or $\gamma' = \cosh\theta\, e_1 + \sinh\theta\, e_2$, depending of whether $\gamma'$ is positive or negative. It follows that

$$\varphi' = \frac{\sinh\theta}{\sin\psi} \quad \text{and} \quad \psi' = \cosh\theta$$

$$\left(\text{resp. } \varphi' = \frac{\cosh\theta}{\sin\psi} \quad \text{and} \quad \psi' = \sinh\theta\right).$$

With the following choice of unit normal vector $\nu := \cosh\theta\, e_1 + \sinh\theta\, e_2$, (resp. $\nu := \sinh\theta\, e_1 + \cosh\theta\, e_2$, we obtain

$$E = -\sinh\theta \sin\psi \cos^{n-1}\psi$$

$$(\text{resp. } E = -\cosh\theta \sin\psi \cos^{n-1}\psi).$$

In the positive case, $\psi$ is monotone and ranges the open interval $(0, \pi/2)$. Due to the symmetry of the problem, we may assume that $\theta > 0$, and the projection of $\gamma$ on the plane $(\theta, \psi)$ has two unbounded ends $(\theta, \psi) \rightarrow$

$(\infty, 0)$ and $(\theta, \psi) \rightarrow (\infty, \pi/2)$, at which the total variation of $\varphi$ is infinite. Therefore the solutions $\gamma$ are embedded curves with two unbounded ends.

In the negative case, the profile of the projections of the solutions $(\gamma, \gamma')$ on the plane $(\theta, \psi)$ are analogous to the spherical case (Section 3.3.4). There is: (i) a fixed point $(0, \psi_0)$, where $\psi_0 := \arccos(\sqrt{\frac{n-1}{n}})$. Its energy level is $E_0 := e(\psi_0)$. The corresponding $\gamma$ has constant curvature; this is the isoparametric case discussed in Section 3.3.3; and (ii) a one-parameter family of closed curves winding around the fixed point $(0, \psi_0)$. However, unlike the spherical case, there is no period problem to study here: since the coordinate $\varphi$ is a hyperbolic angle and is strictly monotone and periodic, its total variation is again infinite and the solutions $\gamma$ are embedded curves with two unbounded ends.

### 3.3.9 *Conclusion*

**Theorem 15.** *Let $S$ be a connected, minimal hypersurface of the space form $\mathbb{Q}_{p,1}^{n+1}$ which is $SO(p', n - p')$-equivariant, with $p' \in \{p, p - 1, p - 2\}$. Then $S$ is congruent to an open subset of one the following*

(i) *a totally geodesic hypersurface $\mathcal{Q}^0$ i.e. the intersection of $\mathbb{Q}_{p,1}^{n+1}$ with a linear hyperplane of $\mathbb{R}^{n+2}$, as described in Theorem 14;*

(ii) *the isoparametric Cartesian products*

$$\sqrt{\frac{1}{n}} \mathbb{Q}_{p,1}^{n-1} \times \sqrt{\frac{n-1}{n}} \{(\cos t, \sin t) | t \in \mathbb{R}/2\pi\mathbb{Z}\} \subset (\mathbb{R}^n, \langle ., . \rangle_p) \oplus (\mathbb{R}^2, \langle ., . \rangle_0),$$

*or*

$$\sqrt{\frac{1}{n}} \mathbb{Q}_{p-1,1}^{n-1} \times \sqrt{\frac{n-1}{n}} \{(\sinh t, \cosh t) | t \in \mathbb{R}\} \subset (\mathbb{R}^n, \langle ., . \rangle_{p-1}) \oplus (\mathbb{R}^2, \langle ., . \rangle_1);$$

(iii) *a generalized catenoid*

$$\{(rx, z_1, z_2) \in \mathbb{Q}_{p,1}^{n+1} | x \in \mathbb{Q}_{p',\epsilon}^{n-1}, \gamma = (r, z_1, z_2) \in \Gamma\},$$

*where $\Gamma := \gamma(\mathbb{R})$ is the image of a curve $\gamma$ of the surface $\Sigma := \mathbb{Q}_{p'',1}^2 \cap \{r > 0\}$, with $p'' = p - p' - \frac{\epsilon + 1}{2}$, endowed with the metric $\epsilon dr^2 + \epsilon' dz_1^2 + \epsilon'' z_2^2$ of signature $(p'', 2 - p'')$. The curve $\gamma$, which has curvature $\kappa$ with respect to the unit normal vector $\nu = (\nu_1, \nu_2, \nu_3)$, is solution of the equation*

$$\kappa - (n - 1)\frac{\nu_1}{r} = 0. \qquad (3.6)$$

*In the spherical case $\Sigma = \mathbb{S}^2 \cap \{r > 0\}$, this equation has a countable family of closed solutions, all of them with cross intersections. The* *corresponding generalized catenoids are therefore never embedded. In the Riemannian case $p = 0$, these hypersurfaces are moreover compact. The other spherical solutions of Equation (3.6) are not properly immersed in $\Sigma$ and their images are dense in an open subset of $\Sigma$. The corresponding catenoids have a dense image in $\mathbb{Q}_{p,1}^{n+1}$ as well.*

*In all the other cases, Equation (3.6) has a one-parameter family of solutions which are complete and embedded curves of $\Sigma$, with two ends which are either a null end-point or an unbounded end. Therefore the corresponding generalized catenoids are complete and embedded hypersurfaces of $\mathbb{Q}_{p,1}^{n+1}$.*

## Remark 18.

(i) When $p' = 0$, the ambient manifold is the sphere $\mathbb{S}^{n+1}$, and only the case $(\epsilon, \epsilon', \epsilon'') = (1, 1, 1)$ occurs; the corresponding hypersurfaces are $SO(n)$-equivariant and therefore foliated by $(n-1)$-spheres; the generating curves are spherical curves. This case has been treated for the first time in [Otsuki (1970)] (cf also [Furuya (1971)], [Brito, Leite (1990)]).

(ii) In the case of the hyperbolic space $\mathbb{H}^n$, there are two families of minimal equivariant hypersurfaces. If $(\epsilon, \epsilon', \epsilon'') = (1, -1, -1)$, the corresponding hypersurfaces are $SO(n-1, 1)$-equivariant and foliated by $(n-1)$-dimensional hyperbolic spaces. If $(\epsilon, \epsilon', \epsilon'') = (-1, -1, 1)$, the corresponding hypersurfaces are $SO(n)$-equivariant and foliated by $(n-1)$-spheres; in both cases the generating curves are hyperbolic curves and of course the induced metric is definite.

(iii) In the case of the de Sitter space $d\mathbb{S}^{n+1}$, there are seven families of solutions. If $(\epsilon, \epsilon', \epsilon'') = (1, 1, 1)$, the corresponding hypersurfaces are $SO(1, n-1)$-equivariant and foliated by $(n-1)$-dimensional de Sitter-type quadrics while the generating curve is spherical and the induced metric is Lorentzian. The cases $(\epsilon, \epsilon', \epsilon'') = (1, -1, -1)$ and $(-1, -1, 1)$ do not occur. If $(\epsilon, \epsilon', \epsilon'') = (-1, 1, 1)$, the corresponding hypersurfaces are $SO(1, n-1)$-equivariant and foliated by $(n-1)$-hyperbolic spaces. The generating curve lies in the de Sitter surface $d\mathbb{S}^2$ and may be positive or negative. In the first case, the induced metric on the corresponding hypersurface is definite, while it is Lorentzian in the second one. Finally, if $(\epsilon, \epsilon', \epsilon'') = (1, -1, 1)$, the corresponding hypersurfaces are $SO(n)$-equivariant and foliated by $(n-1)$-spheres. There are four types of generating curves, lying in the de Sitter surface $d\mathbb{S}^2$, two positives or two negatives. Again, the induced metric is definite or Lorentzian depending or whether the generating curve is positive or negative.

(iv) We recall that $d\mathbb{S}^3$ and $Ad\mathbb{S}^3$ are anti-isometric and therefore have the same geometry. For $n > 2$, there are eight families of solutions in the anti de Sitter space $Ad\mathbb{S}^{n+1}$, four of them with Lorentzian induced metric and four of them with definite induced metric.

(v) For $n > 3$ and $p \notin \{0, 1, n-1, n\}$ all the possible cases do occur and there are nine families of solutions. This is for example the case for the quadric $\mathbb{Q}_2^4$ which has neutral metric with signature $(2,2)$.

## 3.4 Exercises

(1) The *Ricci curvature* of a pseudo-Riemannian metric $g$ is defined to be the trace of its curvature tensor, i.e.

$$Ric(X, Y) = \sum \epsilon_i g(R(e_i, X)Y, e_i),$$

where $e_i$ is an orthonormal basis with $g(e_i, e_i) = \epsilon_i$. The metric $g$ is said to be *Einstein* if there exists a constant $\lambda$ such that $Ric = \lambda g$. Prove that the induced metric of $\mathbb{Q}_{p,1}^n$ is Einstein (use Proposition 7).

(2) Let $\epsilon_1, \epsilon_2 = 1$ or $-1$ and $n_1$ and $n_2$ two integers such that $n_1 + n_2 = n - 1$. Show that the map

$$f : I \times \mathbb{Q}_{p_1,\epsilon_1}^{n_1} \times \mathbb{Q}_{p_2,\epsilon_2}^{n_2} \rightarrow \mathbb{R}^{n_1+1} \oplus \mathbb{R}^{n_2+1} = \mathbb{R}^{n+1}$$
$$(s, x, y) \mapsto (\gamma_1(s)x, \gamma_2(s)y),$$

where $\gamma(s) := (\gamma_1(s), \gamma_2(s))$ is a regular curve in the quadrant $\{(\gamma_1, \gamma_2) \in \mathbb{R}^2 \mid \gamma_1 > 0, \gamma_2 > 0\}$ is a non-degenerate immersion with respect to the pseudo-Riemannian metric $\langle ., . \rangle_{p_1+p_2} = \langle ., . \rangle_{p_1} \oplus \langle ., . \rangle_{p_2}$ of $\mathbb{R}^n$. Write the condition of $\gamma$ insuring that $f$ is minimal with respect to $\langle ., . \rangle_{p_1+p_2}$. These examples were studied in the Riemannian case in [Alencar (1993)] and [Alencar, Barros, Palmas, Reyes, Santos (2005)].

(3) (Polar hypersurfaces) Let $f$ be an immersion of an $n$-dimensional, orientable manifold $\tilde{S}$ in $\mathbb{Q}_{p,1}^{n+1}$ and $N$ a unit normal field along $S := f(\tilde{S})$. Set $\epsilon := |N|_p^2$. The smooth map $\bar{f} : \tilde{S} \rightarrow \mathbb{Q}_{p,\epsilon}^{n+1}$ defined by $\bar{f} = N \circ f^{-1}$ is called *the polar* of $f$.

   (i) Prove that $\bar{f}$ is an immersion if and only if the second fundamental form $h_f$ of $f$ has maximal rank. From now on we assume that it is the case and we set $\bar{S} := \bar{f}(\tilde{S})$;

   (ii) Prove that the polar of the polar of $f$ is $f$ itself;

(iii) Assume that $n = 2$. Prove that $\bar{S}$ is minimal if and only if $S$ is minimal; show by an example that it is not true in higher dimension;

(iv) Prove that if $f$ is isoparametric, so is $\bar{f}$ (see Exercise 2, Chapter 1);

(v) Prove that if $f$ is austere, so is $\bar{f}$ (see Exercise 2, Chapter 1).

(4) (Parallel hypersurfaces) Using the same notations than in the previous exercise, we set

$$f_\theta := \cos\theta\, f + \sin\theta\, \bar{f} \quad \text{if} \quad \epsilon = 1,$$

$$f_\theta := \cosh\theta\, f + \sinh\theta\, \bar{f} \quad \text{if} \quad \epsilon = -1.$$

(i) Prove that there exists $\theta_0$ such that $f_\theta$ is an immersion $\forall\theta$, $\theta \in (-\theta_0, \theta_0)$, and that if $\bar{f}$ is an immersion, then $f_\theta$ is an immersion $\forall\theta \in \mathbb{R}$;

(ii) Prove that if $f$ is austere (respectively isoparametric), so is $f_\theta$.

# Chapter 4

# Pseudo-Kähler manifolds

In this chapter we describe a special class of pseudo-Riemannian manifolds of even dimension, namely the pseudo-Kähler manifolds. In the positive case, we recover the well-known concept of a *Kähler manifold*. We first describe the pseudo-Kähler structures of the complex space and then give the general definition, which is modeled on it. We then introduce the most natural examples of pseudo-Kähler manifolds, the complex pseudo-Riemannian space forms. We conclude the chapter with a more elaborate construction: we show that the tangent bundle of a pseudo-Kähler manifold enjoys itself a pseudo-Kähler structure. In Chapter 5 we shall discuss several classes of minimal submanifolds in any of these types of pseudo-Kähler manifolds.

## 4.1  The complex pseudo-Euclidean space

Consider the complex space $\mathbb{C}^n$ with coordinates $\{z_j = x_j + iy_j, 1 \leq j \leq n\}$. This space may be identified in the obvious way with the real space $\mathbb{R}^{2n}$. The *complex structure $J$* of $\mathbb{C}^n$ is the endomorphism obtained by multiplying all the coefficients of a vectors by the scalar $i$:

$$J(z_1, ..., z_n) := (iz_1, ..., iz_n).$$

In real notations, we have

$$J(x_1, y_1, ..., x_n, y_n) = (-y_1, x_1, ..., -y_n, x_n).$$

The most important property of $J$ is that its square is minus the identity: $J^2 = -Id$.

Next introduce the "pseudo-Hermitian form of signature $(p, n - p)$":

$$\langle\langle \cdot, \cdot \rangle\rangle_p := -\sum_{j=1}^{p} dz_j d\bar{z}_j + \sum_{j=p+1}^{n} dz_j d\bar{z}_j.$$

This is a complex-valued two-form whose real and imaginary parts, hence two real-valued forms, are of great importance:

(i) the two-form

$$\langle \cdot, \cdot \rangle_{2p} := \text{Re} \langle\langle \cdot, \cdot \rangle\rangle_p = - \sum_{j=1}^{p} |dz_j|^2 + \sum_{j=p+1}^{n} |dz_j|^2$$

$$= - \sum_{j=1}^{p} (dx_j^2 + dy_j^2) + \sum_{j=p+1}^{n} (dx_j^2 + dy_j^2)$$

is nothing but the canonical pseudo-Riemannian metric of signature $(2p, 2(n-p))$ in $\mathbb{R}^{2n}$, as considered in the previous chapters;

(ii) the two-form

$$\omega_p := -\text{Im} \langle\langle \cdot, \cdot \rangle\rangle_p = \frac{i}{2} \left( - \sum_{j=1}^{p} dz_j \wedge d\bar{z}_j + \sum_{j=p+1}^{n} dz_j \wedge d\bar{z}_j \right)$$

$$= - \sum_{j=1}^{p} dx_j \wedge dy_j + \sum_{j=p+1}^{n} dx_j \wedge dy_j$$

is alternate and closed; such a form is called *symplectic*.

There are several algebraic relations between the complex structure $J$, the metric $g$ and the symplectic form $\omega$:

$$\langle J\cdot, J\cdot \rangle_{2p} = \langle \cdot, \cdot \rangle_{2p} \qquad \omega_p(J\cdot, J\cdot) = \omega_p$$

$$\omega_p = \langle J\cdot, \cdot \rangle_{2p} \qquad \langle \cdot, \cdot \rangle_{2p} = -\omega_p(J\cdot, \cdot).$$

The group of linear transformations of $\mathbb{C}^n$ which preserve the Hermitian product $\langle\langle \cdot, \cdot \rangle\rangle_p$ (hence both the metric and the symplectic form) will be denoted by

$$U(p, n-p) := \{ M \in Gl(\mathbb{C}^n) \big| \langle\langle MX, MY \rangle\rangle_p = \langle\langle X, Y \rangle\rangle_p \}.$$

We end this section introducing an additional structure on $\mathbb{C}^n$ which will be important in the next two chapters. The *holomorphic volume form* of $\mathbb{C}^n$ is the complex-valued $n$ form

$$\Omega := dz_1 \wedge ... \wedge dz_n.$$

In other words, evaluating $\Omega(X_1, ..., X_n)$, where $X_1, ..., X_n$ are $n$ vectors of $\mathbb{C}^n$ is nothing but calculating the determinant of the $n \times n$ matrix whose columns are the (complex) coordinates of the vectors $X_1, ..., X_n$.

## 4.2 The general definition

A *pseudo-Kähler manifold* is a differentiable manifold $(\mathcal{M}, J, g, \omega)$ of even dimension $2n$ equipped with the three following structures, satisfying furthermore a set of algebraic and analytical conditions:

(i) an almost complex structure, i.e. a $(1, 1)$ tensor $J$, satisfying $J^2 = -Id$;
(ii) a non-degenerate, bilinear two-form $g$, i.e. a pseudo-Riemannian metric;
(iii) a non-degenerate, alternate form $\omega$.

The algebraic conditions that we require are the following: first, the three structures just defined must be related by the formula

$$\omega = g(J., .).$$

Hence $\omega$ is completely determined by $J$ and $g$. Since $J^2 = -Id$, this formula is equivalent to

$$g = -\omega(J., .),$$

so similarly $g$ is determined by $J$ and $\omega$. Moreover, since $g$ is non-degenerate, and recalling Remark 1 of Chapter 1, any one of these two formulae shows that $J$ is determined by $g$ and $\omega$. Hence, any two of those three structures determine the last one.

We furthermore require that $g$ is *Hermitian* with respect to $J$, i.e. $J$ is an isometry of $g$:

$$g(J., J.) = g.$$

Clearly it is equivalent to the fact that

$$\omega(J., J.) = \omega.$$

The analytical conditions are:

(i) the almost complex structure $J$ is actually a *complex structure*. This means that there exists an atlas of *holomorphic* charts $\phi_U : U \to \mathbb{C}^n$ on $\mathcal{M}$, i.e. if $U \cap U' \neq \emptyset$, the diffeomorphism $\phi_U \circ \phi_{U'}^{-1}$ is holomorphic in $\mathbb{C}^n$, and $J$ is defined locally by $d\phi_U \circ J = \tilde{J} \circ d\phi_U$, where here $\tilde{J}$ denotes the canonical complex structure of $\mathbb{C}^n$ introduced in Section 4.1; the fact that $\phi_U \circ \phi_{U'}^{-1}$ is holomorphic implies that the definition of $J$ is independent of the local chart $\phi_U$ considered;
(ii) the form $\omega$ defines a *symplectic structure*, i.e. is closed.

Hence a pseudo-Kähler manifold is both a pseudo-Riemannian manifold, a complex manifold and a symplectic manifold. Because of the pseudo-Riemannian structure, all the concepts introduced in Chapter 1 are relevant here and in particular pseudo-Kähler manifolds enjoy minimal submanifolds. In the next chapter we shall study the minimality condition for some classes of submanifolds which have a special behaviour with respect to the two other structures, the complex and the symplectic ones.

It may be sometimes difficult to check practically whether condition (i) above is satisfied or not. Fortunately, given an almost complex structure, there is a simple property, involving the bracket of vector fields, which turns out to be equivalent to the fact that $J$ is a complex structure. This difficult theorem has been proved in [Newlander, Nirenberg (1957)] (cf also [Hörmander (1990)]):

**Theorem 16.** *An almost complex structure $J$ is actually complex if and only if the following quantity, called the* Nijenhuis tensor, *vanishes:*

$$N^J(X,Y) := [X,Y] + J[JX,Y] + J[X,JY] - [JX,JY].$$

We have seen in Chapter 1 that a pseudo-Riemannian structure admits a canonical connection, the Levi-Civita connection. The next theorem provides a simple criterion to check whether a manifold is pseudo-Kähler:

**Theorem 17.** *Let $\mathcal{M}$ be a manifold equipped with an almost complex structure $J$ and a pseudo-Riemannian metric $g$ which is Hermitian with respect to $J$. Then, setting $\omega := g(J.,.)$, the quadruple $(\mathcal{M}, J, g, \omega)$ is a pseudo-Kähler manifold if and only if $J$ is parallel with respect to the Levi-Civita connection of $g$, i.e. $\nabla_X J = 0, \forall X \in T\mathcal{M}$.*

The proof of this theorem requires two technical lemmas:

**Lemma 9.** *Given a vector field $X$, the operator $Y \mapsto (\nabla_X J)Y$ is skew-symmetric with respect to the metric $g$, i.e. the following formula holds:*

$$g((\nabla_X J)Y, Z) + g(Z, (\nabla_X J)Y) = 0.$$

*Moreover, the operators $J$ and $Y \mapsto (\nabla_X J)Y$ anti-commute, i.e.*

$$J(\nabla_X J)Y + (\nabla_X J)JY = 0.$$

*Proof.* Fix a point $x$ of $\mathcal{M}$ and two vectors $X, Y \in T_x\mathcal{M}$. By an argument similar to that of Lemma 2, Chapter 1, we can extend $X$ and $Y$ in a neighbourhood of $x$ to vector fields, denoted again by $X$ and $Y$ such

$\nabla_X Y(x) = 0$. We first use the fact that $J$ is skew-symmetric, differentiating the identity $g(Y, JY) = 0$ in the direction $X$:

$$g(\nabla_X Y, JY) + g(Y, (\nabla_X J)Y) + g(Y, J(\nabla_X Y)) = g(Y, (\nabla_X J)Y) = 0.$$

Therefore $g(Y + Z, (\nabla_X J)(Y + Z)) = 0$, which gives, using the bilinearity of $g$, the first assertion.

The second assertion follows from a simple calculation. On the one hand we have:

$$J(\nabla_X J)Y = J\big(\nabla_X(JY) - J(\nabla_X Y)\big) = J\nabla_X(JY) + \nabla_X Y,$$

while, on the other hand,

$$(\nabla_X J)(JY) = \nabla_X(J(JY)) - J\nabla_X(JY) = -\nabla_X Y - J\nabla_X(JY).$$

Hence $J$ and $\nabla_X J$ anti-commute. $\qquad\square$

**Lemma 10.** *The Nijenhuis tensor $N^J$ vanishes if and only if*

$$(\nabla_{JX} J)Y = J(\nabla_X J)Y, \forall X, Y \in T\mathcal{M}. \tag{4.1}$$

*Proof.* Fix a point $x$ of $\mathcal{M}$ and two vector fields $X$ and $Y$ such $\nabla_X Y(x)$ vanishes. In particular $\nabla_Y(JX)(x) = (\nabla_Y J)X(x)$. Therefore, we have

$$\begin{aligned}
N^J(X, Y) &= [X, Y] + J[JX, Y] + J[X, JY] - [JX, JY] \\
&= -J(\nabla_Y J)X + J(\nabla_X J)Y - (\nabla_{JX} J)Y + (\nabla_{JY} J)X \\
&= -\big((\nabla_{JX} J)Y - J(\nabla_X J)Y\big) + \big((\nabla_{JY} J)X - J(\nabla_Y J)X\big).
\end{aligned}$$

Hence, if Equation (4.1) holds, $N^J$ must vanish. Conversely, suppose that $N^J$ vanishes and introduce the tri-linear form $T_1(X, Y, Z) := g((\nabla_{JX} J)Y, Z) - g(J(\nabla_X J)Y, Z)$. From the computation above we deduce that the vanishing $N^J$ is equivalent to

$$T_1(X, Y, Z) = T_1(Y, X, Z). \tag{4.2}$$

On the other hand, by Lemma 9,

$$T_1(X, Y, Z) + T_1(X, Z, Y) = 0. \tag{4.3}$$

The combination of Equations (4.2) and (4.3) gives:

$$T_1(X, Y, Z) = -T_1(Y, Z, X). \tag{4.4}$$

Therefore, iterating the circular permutation of entries, we get

$$T_1(X, Y, Z) = -T_1(Y, Z, X) = T_1(Z, X, Y) = -T_1(X, Y, Z),$$

hence $T_1$ vanishes identically, i.e. Equation (4.1) holds. $\qquad\square$

*Proof of Theorem 17.* Assume first that $J$ is parallel. By Lemma 10, it follows that $N^J$ vanishes, and therefore $(\mathcal{M}, J)$ is complex. Moreover, from the identity $\omega = g(J., .)$ and the fact that $g$ is parallel we obtain that $\omega$ is parallel as well, so in particular closed. Therefore $(\mathcal{M}, J, g, \omega)$ is a pseudo-Kähler manifold. Conversely, assume that $(\mathcal{M}, J, g, \omega)$ is pseudo-Kähler. The fact that $J$ is parallel with respect to $g$ is equivalent to the fact that the tri-linear form $T_2(X, Y, Z) := g((\nabla_X J)Y, Z)$ vanishes. To check that this is indeed the case, we first observe that, by Lemma 9,

$$T_2(X, Y, JZ) = T_2(X, JY, Z). \tag{4.5}$$

Also, we have, by Equation (4.1),

$$T_2(JX, Y, Z) + T_2(X, Y, JZ) = 0. \tag{4.6}$$

Therefore, $T_2(JX, Y, Z) + T_2(X, JY, Z)$ vanishes. To conclude, we consider again three vector fields $X$, $Y$ and $Z$ whose mutual covariant derivatives all vanish at $x$, and we use the fact that $\omega$ is closed, getting

$$\begin{aligned}
0 = d\omega(X, Y, Z) &= X\big(\omega(Y, Z)\big) - Y\big(\omega(X, Z)\big) + Z\big(\omega(X, Y)\big) \\
&\quad - \omega([X, Y], Z) + \omega([X, Z], Y) - \omega([Y, Z], X) \\
&= X\big(g(JY, Z)\big) - Y\big(g(JX, Z)\big) + Z\big(g(JX, Y)\big) \\
&= T_2(X, Y, Z) - T_2(Y, X, Z) + T_2(Z, X, Y).
\end{aligned}$$

Applying this formula to the triples $(X, Y, JZ)$ and $(X, JY, Z)$, we get

$$T_2(X, Y, JZ) - T_2(Y, X, JZ) + T_2(JZ, X, Y) = 0$$

and

$$T_2(X, JY, Z) - T_2(JY, X, Z) + T_2(Z, X, JY) = 0.$$

Summing these two equations and taking Equations (4.5) and (4.6) into account, we get that $2T_2(X, Y, JZ)$ vanishes, and therefore $J$ is parallel. $\square$

Of course the simplest example of pseudo-Kähler structure is the quadruple $(\mathbb{C}^n, J, \langle., .\rangle_p, \omega_p)$ described in the previous section. Another simple but important example is the following:

**Proposition 9.** *Let $(\Sigma, g)$ be an oriented Riemannian surface. Denote by $j$ the rotation of angle $\pi/2$ in $T\Sigma$ and set $\omega := g(j., .)$. Then the quadruple $(\Sigma, j, g, \omega)$ is a pseudo-Kähler manifold.*

*Proof.* It follows from the existence of isothermic coordinates (Theorem 6 of Chapter 2): suppose we are given isothermic coordinates $(s, t)$ such that the basis $(\partial_s, \partial_t)$ is positively oriented. Then $j$ is defined by $j\partial_s = \partial_t$ and $j\partial_t = -\partial_s$. Hence the corresponding chart $\phi_U : U \to \Sigma$, where $U$ is an open subset of $\mathbb{R}^2 \simeq \mathbb{C}$ satisfies $d\phi \circ J = j \circ d\phi$. Therefore, given two systems of local coordinates $(s, t)$ and $(s', t')$ and $\phi_U : U \to \Sigma$ and $\phi_{U'} : U' \to \Sigma$, the map $\phi_U \circ \phi_{U'}^{-1}$ is holomorphic. Moreover, since $\Sigma$ is two-dimensional, any alternated two-form is closed, hence the form $\omega = g(j., .)$, is symplectic and the proof is complete. $\qquad\square$

**Remark 19.** A more direct but lengthier proof of Theorem 9 consists of using Theorem 17 and checking that $j$ is parallel, i.e. $\nabla j = 0$. Since the intermediary calculation will be useful later, we give some detail here: using Koszul formula, we find that

$$\nabla_{\partial_s}\partial_s = r_s\partial_s - r_t\partial_t,$$
$$\nabla_{\partial_t}\partial_s = \nabla_{\partial_s}\partial_t = r_t\partial_s + r_s\partial_t,$$
$$\nabla_{\partial_t}\partial_t = -r_s\partial_s + r_t\partial_t.$$

Therefore, we get

$$\nabla_{\partial_s}j\partial_s = j\nabla_{\partial_s}\partial_s, \qquad \nabla_{\partial_s}j\partial_t = j\nabla_{\partial_s}\partial_t,$$

$$\nabla_{\partial_t}j\partial_s = j\nabla_{\partial_t}\partial_s, \qquad \nabla_{\partial_t}j\partial_t = j\nabla_{\partial_t}\partial_t.$$

By linearity, it follows that $\nabla_X jY = j\nabla_X Y$, i.e. $\nabla j = 0$.

The remainder of the chapter is devoted to the description of two other examples pseudo-Kähler manifolds.

## 4.3 Complex space forms

We recall that $J$ denotes the complex structure of the complex Euclidean space $\mathbb{C}^{n+1}$. The set of complex lines of $\mathbb{C}^{n+1}$ is denoted by $\mathbb{CP}^n$ and called *complex projective space*. In "real" terms, the complex lines are simply the two-dimensional linear subspaces of $\mathbb{C}^{n+1}$ which are $J$-invariant. It is a classical result that $\mathbb{CP}^n$ admits a canonical complex structure (see [Moroianu (2007)]). We shall see a proof of this fact in a moment.

Consider now the pseudo-Hermitian form $\langle\langle ., .\rangle\rangle_p$ of signature $(p, n + 1 - p)$ introduced in Section 4.1. As $J$ is an isometry for the metric $\langle ., .\rangle_{2p}$, the induced metric on a complex line $span(X, JX)$ is either positive, negative or totally null. We denote by $\mathbb{CP}^n_p$ (resp. $\mathbb{CP}^n_{p,-}$, $\mathbb{CP}^n_{p,0}$) the sets of positive (resp. negative, resp. null) complex lines of $\mathbb{C}^{n+1}$. We have of course $\mathbb{CP}_0^n = \mathbb{CP}^n$ and $\mathbb{CP}_{0,-}^n = \emptyset$. Since the transformation of $\mathbb{C}^{n+1}$ $\tau : (z_1, ..., z_p, z_{p+1}, ..., z_{n+1}) \mapsto (z_{p+1}, ..., z_{n+1}, z_1, ..., z_p)$ induces an anti-isometry of $\mathbb{CP}_{n+1-p,-}^n$ onto $\mathbb{CP}_p^n$, we shall restrict our study to the set of positive lines. In particular, we shall call $\mathbb{CP}_n^n$ the *complex hyperbolic space* and denote it by $\mathbb{CH}^n$, although the usual (equivalent) definition is $\mathbb{CH}^n = \mathbb{CP}_{1,-}^n$. In analogy with the space forms of Chapter 3, we propose the terminology *complex de Sitter space* for $\mathbb{CP}_1^n$ and *complex anti de Sitter space* for $\mathbb{CP}_{n-1}^n$. We shall show that all these sets enjoy a pseudo-Kähler structure. For this purpose we consider again the space form introduced in Chapter 2:

$$\mathbb{Q}_{2p,1}^{2n+1} := \{z \in \mathbb{C}^{n+1}, \langle z, z \rangle_{2p} = 1\}.$$

For sake of brevity, we set $\mathbb{Q} := \mathbb{Q}_{2p,1}^{2n+1}$. Observe that the map $z \mapsto Jz$ defines a tangent vector field on $\mathbb{Q}$ (the *Reeb field*), whose orbits $\theta \mapsto \cos\theta\, z + \sin\theta\, Jz := \exp(J\theta)z$ are closed, and are exactly the intersection of a complex line $span(z, Jz)$ with $\mathbb{Q}$. On the other hand, any element of $\mathbb{CP}_p^n$ intersects $\mathbb{Q}$.

Hence the set $\mathbb{CP}_p^n$ can be identified with the quotient of $\mathbb{Q}$ by the equivalence relation $\sim$ defined by

$$z \sim z' \Leftrightarrow \exists \theta \text{ s.t. } z' = e^{i\theta}z = \cos\theta\, z + \sin\theta\, Jz.$$

We shall denote by $\pi$ the canonical projection (called *Hopf projection*):

$$\pi : \mathbb{Q} \to \mathbb{CP}_p^n.$$

We now introduce a fundamental object: the hyperplane distribution $(Jz)^\perp = d\pi^{-1}(T_{\pi(z)}\mathbb{CP}_p^n)$. These hyperplanes of $T\mathbb{Q}$, which are $J$-invariant (in "complex" terms, they are complex linear hyperspaces of $\mathbb{C}^{n+1}$), will allow us describe the geometry of $\mathbb{CP}_p^n$. Observe that given a vector field $X$ on $\mathbb{CP}_p^n$, there exists a unique vector field $\tilde{X}$ on $\mathbb{Q}$, called *horizontal lift*, such that

$$d\pi(\tilde{X})(z) = X(\pi(z)) \text{ and } \tilde{X}(z) \in (Jz)^\perp, \forall z \in \mathbb{Q}.$$

We are now in position to define three structures on $\mathbb{CP}_p^n$:

(i) an almost complex structure $\mathbb{J}$ defined by

$$\mathbb{J}X := d\pi(J\tilde{X});$$

(ii) a pseudo-Riemannian metric $\mathbb{G}$ defined by

$$\mathbb{G}(X, Y) := \langle \tilde{X}, \tilde{Y} \rangle_{2p+2};$$

(The projection $\pi$ is then said to be *pseudo-Riemannian submersion* from $(\mathbb{Q}, \langle ., . \rangle_{2p+2})$ to $(\mathbb{CP}_p^n, \mathbb{G})$);

(iii) the natural alternated two-form $\varpi := \mathbb{G}(\mathbb{J}., .)$. Of course it is equivalent to set $\varpi(X, Y) = \omega_p(\tilde{X}, \tilde{Y})$.

The set of algebraic conditions that the triple $(\mathbb{J}, \mathbb{G}, \varpi)$ must satisfy is easy to check and left to the reader. Observe furthermore that the signature of $\mathbb{G}$ is $(2p, 2(n - p))$. In particular the classical complex projective space $\mathbb{CP}_0^n$ and the complex hyperbolic space $\mathbb{CH}^n = \mathbb{CP}_n^n$ have a definite metric. In the first case, the metric $\mathbb{G}$ is called *Fubini-Study* metric. We shall give an alternative description of the 2-dimensional spaces $\mathbb{CP}^1$ and $\mathbb{CH}^1$ in Section 4.3.1. The 4-dimensional space $\mathbb{CP}_1^2$ has neutral signature $(2, 2)$.

Our next task is to describe the Levi-Civita connection of $\mathbb{G}$ that we denote by $\nabla^{\mathbb{G}}$. Among other things, this will useful in order to establish that $(\mathbb{CP}_p^n, \mathbb{J}, \mathbb{G}, \varpi)$ is pseudo-Kähler. We denote by $D$ the Levi-Civita connection of the induced metric of $\mathbb{Q}$.

**Proposition 10.** *Let $X$ and $Y$ two tangent vector fields of $\mathbb{CP}_p^n$. We have*

$$D_{\tilde{X}}\tilde{Y} = \widetilde{\nabla^{\mathbb{G}}_X Y} + \varpi(Y, X)Jz. \tag{4.7}$$

*Proof.* Given to tangent vector fields $X$ and $Y$ on $\mathbb{CP}_p^n$, we have

$$\mathbb{G}(X, Y) \circ \pi = \langle \tilde{X}, \tilde{Y} \rangle_{2p+2}.$$

On the other hand we may check that $[\tilde{X}, \tilde{Y}] = \widetilde{[X, Y]}$. Hence, using Koszul formula (see Lemma 1 of Chapter 1) we get:

$$\begin{aligned}
2\mathbb{G}(\nabla^{\mathbb{G}}_X Y, Z) &= X(\mathbb{G}(Y, Z)) + Y(\mathbb{G}(X, Z)) - Z(\mathbb{G}(X, Y)) \\
&\quad + \mathbb{G}([X, Y], Z) - \mathbb{G}([X, Z], Y) - \mathbb{G}([Y, Z], X) \\
&= \tilde{X}(\langle \tilde{Y}, \tilde{Z} \rangle_{2p+2}) + \tilde{Y}(\langle \tilde{X}, \tilde{Z} \rangle_{2p+2}) - \tilde{Z}(\langle \tilde{X}, \tilde{Y} \rangle_{2p+2}) \\
&\quad + \langle [\tilde{X}, \tilde{Y}], \tilde{Z} \rangle_{2p+2} - \langle [\tilde{X}, \tilde{Z}], \tilde{Y} \rangle_{2p+2} - \langle [\tilde{Y}, \tilde{Z}], \tilde{X} \rangle_{2p+2} \\
&= 2\langle D_{\tilde{X}}\tilde{Y}, \tilde{Z} \rangle_{2p+2}.
\end{aligned}$$

It follows that

$$\nabla^{\mathbb{G}}_X Y = d\pi(D_{\tilde{X}}\tilde{Y}). \tag{4.8}$$

It remains to calculate the vertical part of $D_{\tilde{X}}\tilde{Y}$ i.e. its component in the direction $Jz$. For this purpose we differentiate the identity $\langle \tilde{Y}, Jz \rangle_{2p+2} = 0$ in the direction $\tilde{X}$, getting

$$\langle D_{\tilde{X}}\tilde{Y}, Jz \rangle_{2p+2} + \langle \tilde{Y}, J\tilde{X} \rangle_{2p+2} = 0.$$

Hence

$$\langle D_{\tilde{X}}\tilde{Y}, Jz \rangle_{2p+2} = -\langle \tilde{Y}, J\tilde{X} \rangle_{2p+2} = \mathbb{G}(X, \mathbb{J}Y) = \varpi(Y, X). \tag{4.9}$$

Putting Equations (4.8) and (4.9) together, and taking into account that $|Jz|^2_{2p+2} = 1$ yield Equation (4.7). $\qquad\Box$

**Theorem 18.** $(\mathbb{CP}^n_p, \mathbb{J}, \mathbb{G}, \varpi)$ *is a pseudo-Kähler manifold.*

*Proof.* According to Theorem 17, it is sufficient to prove that, given two vector fields $X$ and $Y$ of $\mathbb{CP}^n_p$, we have $\nabla_X \mathbb{J}Y = \mathbb{J}\nabla_X Y$. Using Proposition 10, the fact that $J$ is parallel with respect to $D$ and the definition of $\mathbb{J}$, we have

$$\nabla_X \mathbb{J}Y = d\pi(D_{\tilde{X}}\widetilde{\mathbb{J}Y}) = d\pi(D_{\tilde{X}} J\tilde{Y})$$
$$= d\pi(J D_{\tilde{X}}\tilde{Y}) = \mathbb{J}d\pi(D_{\tilde{X}}\tilde{Y}) = \mathbb{J}\nabla_X Y. \qquad\Box$$

**Theorem 19.** *The curvature tensor $R^\nabla$ of the Levi-Civita connection $\nabla^\mathbb{G}$ of $\mathbb{G}$ is given by the formula*

$$R^\nabla(X,Y)Z = \mathbb{G}(X,Z)Y - \mathbb{G}(Y,Z)X + \mathbb{G}(\mathbb{J}X,Z)\mathbb{J}Y - \mathbb{G}(\mathbb{J}Y,Z)\mathbb{J}X. \quad (4.10)$$

*Proof.* Let $X, Y$ and $Z$ three vector fields on $\mathbb{CP}^n_p$. By Equation (4.7) we have

$$D_{\tilde{X}}\tilde{Z} = \widetilde{\nabla^\mathbb{G}_X Z} + \omega(Z,X)Jz = \widetilde{\nabla^\mathbb{G}_X Z} + \omega_p(\tilde{Z},\tilde{X})Jz.$$

It follows that

$$D_{\tilde{Y}}D_{\tilde{X}}\tilde{Z} = D_{\tilde{Y}}\widetilde{\nabla^\mathbb{G}_X Z} + D_{\tilde{Y}}\Big(\omega_p(\tilde{Z},\tilde{X})Jz\Big)$$
$$= \widetilde{\nabla^\mathbb{G}_Y \nabla^\mathbb{G}_X Z} + \omega_p(\nabla^\mathbb{G}_X Z, Y)Jz$$
$$+ \tilde{Y}\big(\omega_p(\tilde{Z},\tilde{X})\big)Jz + \omega_p(\tilde{Z},\tilde{X})J\tilde{Y},$$

hence

$$d\pi(D_{\tilde{Y}}D_{\tilde{X}}\tilde{Z}) = \nabla^\mathbb{G}_Y \nabla^\mathbb{G}_X Z + \varpi(Z,X)\mathbb{J}Y.$$

Analogously

$$d\pi(D_{\tilde{X}}D_{\tilde{Y}}\tilde{Z}) = \nabla^\mathbb{G}_X \nabla^\mathbb{G}_Y Z + \varpi(Z,Y)\mathbb{J}X.$$

Moreover,

$$D_{\widetilde{[X,Y]}}\tilde{Z} = \widetilde{\nabla^\mathbb{G}_{[X,Y]} Z} + \omega_p(\tilde{Z},\widetilde{[X,Y]})Jz.$$

Using the fact that

$$d\pi(R^D(\tilde{X},\tilde{Y})\tilde{Z}) = d\pi(D_{\tilde{Y}}D_{\tilde{X}}\tilde{Z}) - d\pi(D_{\tilde{X}}D_{\tilde{Y}}\tilde{Z}) + d\pi(D_{[\tilde{X},\tilde{Y}]}\tilde{Z}),$$

we deduce that

$$R^\nabla(X,Y)Z = \nabla^\mathbb{G}_Y \nabla^\mathbb{G}_X Z - \nabla^\mathbb{G}_X \nabla^\mathbb{G}_Y Z + \nabla^\mathbb{G}_{[X,Y]} Z$$
$$= d\pi(R^D(\tilde{X},\tilde{Y})\tilde{Z}) + \varpi(Z,Y)\mathbb{J}X - \varpi(Z,X)\mathbb{J}Y.$$

On the other hand, we have seen in Chapter 3 (Proposition 7) that

$$R^D(\tilde{X}, \tilde{Y})\tilde{Z} = \langle \tilde{X}, \tilde{Z} \rangle_{2p+2} \tilde{Y} - \langle \tilde{Y}, \tilde{Z} \rangle_{2p+2} \tilde{X},$$

so that

$$d\pi(R^D(\tilde{X}, \tilde{Y})\tilde{Z}) = \mathbb{G}(X, Z)Y - \mathbb{G}(Y, Z)X.$$

It follows that

$$R^\nabla(X, Y)Z = \mathbb{G}(X, Z)Y - \mathbb{G}(Y, Z)X + \mathbb{G}(\mathbb{J}Z, Y)\mathbb{J}X - \mathbb{G}(\mathbb{J}Z, X)\mathbb{J}Y,$$

which is the required formula. $\qquad\square$

### 4.3.1 *The case of dimension $n = 1$*

It turns out that the two complex space forms of (complex) dimension one, i.e. $\mathbb{CP}^1$ and $\mathbb{CH}^1$, enjoy an alternative description, quite more explicit: they actually are isometric, up to scaling, to the two definite space forms of (real) dimension two, i.e. the sphere and the hyperbolic plane. This fact will be very useful in the next chapter. In order to deal simultaneously with the two cases, we adopt a notation similar to that of Section 3.3 of Chapter 3 and let $\epsilon$ to be 1 or $-1$, set $p = 1 - \epsilon$ and consider the quadric

$$\mathbb{Q}^2_{p,1/4} := \left\{ (x_1, x_2, x_3) \,\middle|\, x_1^2 + \epsilon(x_2^2 + x_3^2) = \frac{1}{4} \right\},$$

equipped with the orientation induced from the canonical orientation of $\mathbb{R}^3$. It is not difficult to check that the corresponding complex structure $j$, (see Theorem 9 in Section 2 of this chapter) takes the following coordinate-free form: $jX := \epsilon\, 2x \times X$, where $X$ is a tangent vector to $\mathbb{Q}^2_{p,1/4}$ at the point $x$, and $\times$ denotes the exterior product (vectorial product) of $\mathbb{R}^3$, defined by the formula: $\langle u \times v, w \rangle_{1-\epsilon} = \det(u, v, w)$.

Moreover, we adopt for brevity the notation $x_1 + ix_2$. Finally, we introduce the map

$$\begin{aligned} \pi_\epsilon: \quad \mathbb{Q}^3_{p,1} &\rightarrow \mathbb{Q}^2_{p,1/4} \\ (z_1, z_2) &\mapsto \frac{1}{2}\Big( (-\epsilon|z_1|^2 + |z_2|^2), 2z_1\bar{z}_2 \Big). \end{aligned}$$

Since $\pi_\epsilon(e^{it}z_1, e^{it}z_2) = \pi_\epsilon(z_1, z_2)$ and $\pi_\epsilon$ is onto, the map $\pi_\epsilon \circ \pi^{-1} : \mathbb{CP}^1_p \rightarrow \mathbb{Q}^2_{p,1/4}$ is well defined and one to one. Moreover, given a tangent vector $X = (X_1, X_2)$ on $\mathbb{Q}^3_{p,1}$ at the point $z = (z_1, z_2)$, we have $\langle X, z \rangle_p = 0$. If $X$

is horizontal, we also have $\langle X, Jz \rangle_p = 0$, and hence $\langle\langle X, z \rangle\rangle_p = \epsilon X_1 \bar{z}_1 + X_2 \bar{z}_2 = 0$. It follows that

$$
d\pi_\epsilon(X_1, X_2) = \frac{1}{2}\Big( (-\epsilon \bar{z}_1 \bar{X}_1 - \epsilon \bar{z}_1 X_1 + z_2 \bar{X}_2 + \bar{z}_2 X_2), 2(z_1 \bar{X}_2 + X_1 \bar{z}_2) \Big)
$$

$$
= \Big( -\epsilon \bar{z}_1 \bar{X}_1 + z_2 \bar{X}_2, z_1 \bar{X}_2 + X_1 \bar{z}_2 \Big).
$$

Therefore, using the fact that $\epsilon |z_1|^2 + |z_2|^2 = 1$, we get

$$
|d\pi_\epsilon(X)|_p^2 = \epsilon |X_1|^2 + |X_2|^2 = |X|_p^2 = \mathbb{G}(d\pi(X), d\pi(X)).
$$

By the polarization formula $2\langle X, Y \rangle_p = |X+Y|_p^2 - |X|_p^2 - |Y|_p^2$, it follows that $\pi_\epsilon \circ \pi^{-1}$ is an isometry between $(\mathbb{CP}^1, \mathbb{G})$ and $(\frac{1}{2}\mathbb{S}^2, \langle ., . \rangle_0)$ on the one hand, and between $(\mathbb{CH}^1, \mathbb{G})$ on $(\frac{1}{2}\mathbb{H}^2, \langle ., . \rangle_2)$ on the other hand.

Finally, since $d\pi_\epsilon(\mathbb{J}X)$ is orthogonal to $d\pi_\epsilon(X)$ and $T_{\pi(z)}\mathbb{Q}_{p,1/4}^2$ is 2-dimensional, we have $d\pi_\epsilon(\mathbb{J}X) = \pm j d\pi_\epsilon(X)$. It remains to check that we have actually $d\pi_\epsilon(\mathbb{J}X) = j d\pi_\epsilon(X)$. The tangent bundle $T\mathbb{Q}_{p,1}^3$ is simply connected, so by a continuity argument, it is sufficient to check that this is true at a given pair $(z, X)$ of $T\mathbb{Q}_{p,1}^3$, for example $z = (1,0)$ and $X = (0,1)$. This is a straightforward computation.

## 4.4   The tangent bundle of a pseudo-Kähler manifold

In this section, we describe the construction, recently uncovered in [Guilfoyle, Klingenberg (2005)], [Guilfoyle, Klingenberg (2008)] (see also [Anciaux, Guilfoyle, Romon (2009)]), of a pseudo-Kähler structure on the tangent bundle of a manifold $\mathcal{M}$ which is equipped itself with a pseudo-Kähler metric. We shall proceed in several steps, underlying at each one which structure of $\mathcal{M}$ is necessary to get a given structure on $T\mathcal{M}$. Along the process, we shall also describe the Sasaki metric, which is canonically defined on the tangent bundle of a pseudo-Riemannian manifold.

### 4.4.1   *The canonical symplectic structure of the cotangent bundle $T^*\mathcal{M}$*

It is a very important fact that given any differentiable manifold $\mathcal{M}$, its cotangent bundle $T^*\mathcal{M}$ enjoys a canonical symplectic structure. We insist on the fact that this structure depends only on the differentiable structure of $\mathcal{M}$.

We first define on $T^*\mathcal{M}$ the canonical *Liouville form* $\alpha$ as follows: let $\pi^* : T^*\mathcal{M} \to \mathcal{M}, (x, \xi) \mapsto x$ the canonical projection. Its differential $d\pi^*$

maps $TT^*\mathcal{M}$ to $T\mathcal{M}$. For a tangent vector $X$ at some point $(x, \xi)$ of $T^*\mathcal{M}$, we set

$$\alpha(X) := \xi(d\pi(X)).$$

The only difficult issue is to check that the object defined by the equation above is actually a one-form on $T^*\mathcal{M}$! We now define the canonical symplectic form $\omega^*$ on $T^*\mathcal{M}$ by setting $\omega^* := -d\alpha$. This is thus a closed 2-form, and to prove that it actually defines a symplectic structure on $T^*\mathcal{M}$, we only have to prove that it is non-degenerate. The simplest way to do so is introducing coordinates on $\mathcal{M}$, which shall prove useful for other purposes later. So, considering local coordinates $(q_1, ..., q_n)$ on $\mathcal{M}$, a one-form takes the following form: $\xi(x) = \sum_{k=1}^n p_k(x) dq_k$, where $p_1, ..., p_n$ are $n$ real functions. Hence $(q_1, ..., q_n, p_1, ..., p_n)$ are natural local coordinates on $T^*\mathcal{M}$. Now, given a tangent vector

$$X = \sum_{i=1}^n X_i \frac{\partial}{\partial q_i} + \sum_{i=1}^n Y_i \frac{\partial}{\partial p_i},$$

we have

$$d\pi^*(X) = \sum_{i=1}^n X_i \frac{\partial}{\partial q_i},$$

so that $\alpha(X) = \sum_{i=1}^n X_i p_i$. It follows that

$$\alpha = \sum_{i=1}^n p_i dq_i$$

and

$$\omega^* = \sum_{i=1}^n dq_i \wedge dp_i,$$

which is clearly non degenerate.

**Remark 20.** It can be actually proved that *any* symplectic manifold $(\mathcal{M}, \omega)$ enjoys local coordinates $(q_1, ..., q_n, p_1, ..., p_n)$ such that the symplectic form writes $\omega = \sum_{i=1}^n dq_i \wedge dp_i$. This result is known as *Darboux theorem* and can be restated as follows: there is no local symplectic invariant.

### 4.4.2 *An almost complex structure on the tangent bundle* $T\mathcal{M}$ *of a manifold equipped with an affine connection*

We now consider the tangent bundle $T\mathcal{M}$, and denote by $\pi$ the canonical projection $T\mathcal{M} \to \mathcal{M}, (x, V) \mapsto x$. The subbundle $Ker(d\pi)$ of $TT\mathcal{M}$ (it is thus a bundle over $T\mathcal{M}$) will be called *the vertical bundle* and denoted by $V\mathcal{M}$. However, although it is quite intuitive that the tangent bundle of $T\mathcal{M}$ splits as $TT\mathcal{M} = V\mathcal{M} \oplus H\mathcal{M} \simeq T\mathcal{M} \oplus T\mathcal{M}$, where $H\mathcal{M}$ is some "horizontal" subbundle, there is *a priori* no such canonical object. Nevertheless, it is possible to define a bundle $H\mathcal{M}$ provided we are given an affine connection $\nabla$ on the base manifold $\mathcal{M}$ as follows: let $X$ be a tangent vector to $T\mathcal{M}$ at some point $(x_0, V_0)$. This implies that there exists a curve $\gamma(s) = (x(s), V(s))$ such that $(x(0), V(0)) = (x_0, V_0)$ and $\gamma'(0) = X$. If $X \notin V\mathcal{M}$ (which implies $x'(0) \neq 0$), we define the connection map (see [Kowalski (1971)], [Dombrowski (1962)]) $K : TT\mathcal{M} \to T\mathcal{M}$ by $KX = \nabla_{x'(0)}V(0)$, which does not depend on the curve $\gamma$. If $X$ is vertical, we may assume that the curve $\gamma$ stays in a fiber so that $V(s)$ is a curve in a vector space. We then define $KX$ to be simply $V'(0)$. The horizontal bundle is then $Ker(K)$ and we have a direct sum

$$TT\mathcal{M} = H\mathcal{M} \oplus V\mathcal{M} \simeq T\mathcal{M} \oplus T\mathcal{M}$$
$$X \simeq (\Pi X, KX). \tag{4.11}$$

Here and in the following, $\Pi$ is a shorthand notation for $d\pi$.

Once we have constructed this decomposition, there is a straightforward way to define an almost complex structure on $T\mathcal{M}$: we define it $J_S$ by the following:

$$J_S(\Pi X, KX) = (-KX, \Pi X). \tag{4.12}$$

Unfortunately, it turns out that this almost complex structure is almost never complex!

**Theorem 20.** *Let $(\mathcal{M}, \nabla)$ a manifold endowed with an affine connection. The almost complex structure $J_S$ defined by Equation (4.12) is not complex unless $\nabla$ is flat.*

In order to prove this result, we shall need a technical lemma, which shows how to compute the bracket of vector fields on $T\mathcal{M}$, using the decomposition of $TT\mathcal{M}$:

**Lemma 11.** *[Dombrowski (1962)] Given a vector field $X$ on $(\mathcal{M}, \nabla)$ there exists exactly one vector field $X^h$ and one vector field $X^v$ on $T\mathcal{M}$ such that*

$(\Pi X^h, KX^h) = (X, 0)$ *and* $(\Pi X^v, KX^v) = (0, X)$. *Moreover, given two vector fields* $X$ *and* $Y$ *on* $(\mathcal{M}, \nabla)$, *we have, at the point* $(x, V)$:

$$[X^v, Y^v] = 0,$$
$$[X^h, Y^v] \simeq (0, \nabla_X Y),$$
$$[X^h, Y^h] \simeq ([X, Y], -R(X, Y)V),$$

*where* $R$ *denotes the curvature of* $\nabla$ *and we use the direct sum notation* (4.11).

We say that a vector field $X$ on $T\mathcal{M}$ is *projectable* if it is constant on the fibres. According to the lemma above, it is equivalent to the fact that there exists two vector fields $X_1$ and $X_2$ on $\mathcal{M}$ such that $X = (X_1)^h + (X_2)^v$.

*Proof of Theorem 20.* By Theorem 16, it is sufficient to compute the Nijenhuis tensor of $J_S$: Since $N^{J_S}(X, Y)$ depends pointwise on the tangent vectors we may assume for computational purposes that $X$ and $Y$ are projectable, and use Lemma 11. By linearity and skew-symmetry it suffices to compute $N^{J_S}(X, Y)$ in three distinct cases:

(i) vertical fields

$$\begin{aligned}
N^{J_S}(X^v, Y^v) &= [X^v, Y^v] + J_S[J_S X^v, Y^v] + J_S[X^v, J_S Y^v] - [J_S X^v, J_S Y^v] \\
&= [X^v, Y^v] + J_S[-X^h, Y^v] + J_S[X^v, -Y^h] - [-X^h, -Y^h] \\
&= 0 + J_S(0, -\nabla_X Y) + J_S(0, \nabla_Y X) - ([X, Y], -R(X, Y)V) \\
&= (\nabla_X Y, 0) + (-\nabla_Y X, 0) - ([X, Y], -R(X, Y)V) \\
&= -(0, R(X, Y)V).
\end{aligned}$$

(ii) horizontal fields

$$\begin{aligned}
N^{J_S}(X^h, Y^h) &= [X^h, Y^h] + J_S[J_S X^h, Y^h] + J_S[X^h, J_S Y^h] - [J_S X^h, J_S Y^h] \\
&= ([X, Y], -R(X, Y)V) + J_S(0, -\nabla_Y X) + J_S(0, \nabla_X Y) - 0 \\
&= -(0, R(X, Y)V).
\end{aligned}$$

(iii) mixed fields

$$\begin{aligned}
N^{J_S}(X^h, Y^v) &= [X^h, Y^v] + J_S[J_S X^h, Y^v] + J_S[X^h, J_S Y^v] - [J_S X^h, J_S Y^v] \\
&= (0, \nabla_X Y) + J_S(-[X, Y], R(X, Y)V) - (0, \nabla_Y X) \\
&= -(R(X, Y)V, 0).
\end{aligned}$$

We conclude that $N^{J_S}$ vanishes if and only if $R$ does so, i.e. if $\nabla$ is flat. $\square$

### 4.4.3 *Identifying $T^*\mathcal{M}$ and $T\mathcal{M}$ and the Sasaki metric*

We now equip $\mathcal{M}$ with a pseudo-Riemannian metric $g$. By the non-degeneracy assumption, we can identify $T^*\mathcal{M}$ with $T\mathcal{M}$ by the following isomorphism:

$$\begin{aligned} \iota: T\mathcal{M} &\to T\mathcal{M}^* \\ (x, V) &\mapsto (x, \xi), \end{aligned}$$

where $\xi$ is defined by

$$\xi(W) := g(V, W), \quad \forall W \in T_x\mathcal{M}.$$

It follows that we can transport any structure of $T^*\mathcal{M}$ onto $T\mathcal{M}$ and conversely. For example, pulling back $\omega^*$ by $\iota$, we get a symplectic structure $\omega := \iota^*(\omega^*)$ on $T\mathcal{M}$.

Let us see how this identification writes in coordinates and how we can deduce a coordinate expression of $\omega$: given a local system of coordinates $(x_1, ..., x_n)$ in $\mathcal{M}$, a tangent vector takes the form $V = \sum_{i=1}^{n} x_{n+i}\partial_{x_i}$, so that $(x_1, ..., x_n, x_{n+1}, ..., x_{2n})$ is a natural system of coordinates in $T\mathcal{M}$. Now if $(x_1, ..., x_n) = (q_1, ..., q_n)$, the one-form $\xi = \sum_{i=1}^{n} p_i dq_i$ is identified with the vector $V = \sum_{i=1}^{n} x_{n+i}\partial_{q_i}$ such that $p_i = \sum_{j=1}^{n} g_{ij}x_{n+j}$, or, equivalently, $x_{n+j} = \sum_{i=1}^{n} g^{ij}p_i$ (as usual, $g_{ij}$ are the coefficients of the pseudo-Riemannian metric $g$). It follows that

$$\omega = \iota^*\left(\sum_{i=1}^{n} dq_i \wedge dp_i\right) = \sum_{i,j=1}^{n} g_{ij}dx_i \wedge dx_{n+j}.$$

Thanks to the existence of the Levi-Civita connection of the metric $g$, we may apply the construction of the previous section, and we have a splitting $TT\mathcal{M} = H\mathcal{M} \oplus V\mathcal{M}$. There is a beautiful, explicit formula for $\omega$ in terms of this splitting, which will prove useful in the next chapter:

**Lemma 12.** *Let $X$ and $Y$ be two tangent vectors to $T\mathcal{M}$; we have*

$$\omega(X, Y) = g(KX, \Pi Y) - g(\Pi X, KY).$$

*Proof.* Writing $X = \sum_{i=1}^{2n} X_i\partial_{x_i}$, we have $\Pi X = \sum_{i=1}^{n} X_i\partial_{x_i}$. It is a bit more difficult to express $KX$: by definition, if $\gamma(s) = (x(s), V(s))$, with $V(s) = \sum V_i(s)\partial_{x_i}$, is a parametrized curve in $T\mathcal{M}$ such that $(x'(0), V'(0)) = X$ at some point $(x_0, V_0) = (x(0), V(0))$, then $KX =$

$\nabla_{x'(0)} V(0)$. Therefore

$$
\begin{aligned}
KX &= \sum_{i=1}^{n} \frac{dv_i}{ds} \partial_{x_i} + \sum_{i=1}^{n} v_i \nabla_{x'(0)} \partial_{x_i} \\
&= \sum_{i=1}^{n} X_{i+n} \partial_{x_i} + \sum_{i,j=1}^{n} v_i \frac{dx_j}{ds} \nabla_{\partial_{x_j}} \partial_{x_i} \\
&= \sum_{i=1}^{n} X_{i+n} \partial_{x_i} + \sum_{i,j=1}^{n} v_i X_j \nabla_{\partial_{x_j}} \partial_{x_i}.
\end{aligned}
$$

It follows that, setting $Y = \sum_{i=1}^{2n} Y_i \partial_{x_i}$, we have

$$
g(KX, \Pi Y) = \sum_{i,j=1}^{n} X_{n+i} Y_i g_{ij} + \sum_{i,j,k=1}^{n} v_i X_j Y_k \, g\left(\nabla_{\partial_{x_i}} \partial_{x_j}, \partial_{x_k}\right).
$$

Analogously

$$
g(\Pi X, KY) = \sum_{i,j=1}^{n} Y_{n+i} X_i g_{ij} + \sum_{i,j,k=1}^{n} v_i Y_j X_k \, g\left(\nabla_{\partial_{x_i}} \partial_{x_j}, \partial_{x_k}\right).
$$

We now use Lemma 2 of Chapter 1, that insures the existence of normal coordinates, i.e. such that the quantities $\nabla_{\partial_{x_i}} \partial_{x_j}$ vanish at some point $x$. Therefore, at this point $x$, we have

$$
\begin{aligned}
g(KX, \Pi Y) - g(\Pi X, KY) &= \sum_{i,j=1}^{n} (X_{n+i} Y_j - Y_{n+i} X_j) g_{ij} \\
&= \left( \sum_{i,j=1}^{n} g_{ij} dx_i \wedge dx_{n+j} \right) (X, Y) \\
&= \omega(X, Y). \qquad \square
\end{aligned}
$$

This formula allows us to prove in particular the following:

**Lemma 13.** *The symplectic form $\omega$ of $T\mathcal{M}$ is compatible with the complex structure $J_S$ defined in the previous subsection.*

*Proof.* By Lemma 12, we easily compute:

$$
\begin{aligned}
\omega(J_S X, J_S Y) &= g(K J_S X, \Pi J_S Y) - g(\Pi J_S X, K J_S Y) \\
&= g(\Pi X, -KY) - g(-KX, \Pi Y) \\
&= \omega(X, Y). \qquad \square
\end{aligned}
$$

We now recall that an almost complex structure together with a symplectic form determine a pseudo-Riemannian metric. We shall denote by $G_S$ the metric obtained from $J_S$:

$$G_S := -\omega(J_S., .).$$

Equivalently, $G_S$ may be defined by

$$G_S(X, Y) = g(\Pi X, \Pi Y) + g(KX, KY).$$

It was actually by this natural formula that the metric $G_S$, usually called the *Sasaki metric,* was historically introduced. The Sasaki metric has been extensively studied (see [Boeckx, Vanhecke (2000)] and references therein), probably because of the natural way in which it is defined, although, as Theorem 20 shows, the triple $(J_S, G_S, \omega)$ is not a Kähler structure when $(\mathcal{M}, g)$ is not flat.

### 4.4.4   *A complex structure on the tangent bundle of a pseudo-Kähler manifold*

Finally, we assume that $\mathcal{M}$ is equipped with a pseudo-Kähler structure $(J, g, \varpi)$. In particular, $\mathcal{M}$ must have even dimension and the dimension of $T\mathcal{M}$ is therefore a multiple of four. Thus we still have a splitting of $TT\mathcal{M}$ and we define a complex structure by $\mathbb{J} := J \oplus J$, i.e.

$$\mathbb{J}(\Pi X, KX) = (J\Pi X, JKX).$$

**Lemma 14.** *The symplectic form $\omega$ is compatible with $\mathbb{J}$.*

*Proof.*   Using Lemma 12, we compute

$$\omega(\mathbb{J}X, \mathbb{J}Y) = g(K\mathbb{J}X, \Pi\mathbb{J}Y) - g(\Pi\mathbb{J}X, K\mathbb{J}Y)$$
$$= g(JKX, J\Pi Y) - g(J\Pi X, JKY)$$
$$= \omega(X, Y). \qquad \square$$

**Theorem 21.** $\mathbb{J}$ *is a complex structure.*

*Proof.*   We proceed exactly as in the proof of Theorem 20, proving that $N^{\mathbb{J}}$ vanishes in three distinct cases:

(i) vertical fields

$$N^{\mathbb{J}}(X^v, Y^v) = 0.$$

(ii) horizontal fields

$$N^{\mathbb{J}}(X^h, Y^h) = [X^h, Y^h] + \mathbb{J}[\mathbb{J}X^h, Y^h] + \mathbb{J}[X^h, \mathbb{J}Y^h] - [\mathbb{J}X^h, \mathbb{J}Y^h]$$
$$= ([X,Y], -R(X,Y)V) + \mathbb{J}([JX,Y], -R(JX,Y)V)$$
$$+ \mathbb{J}([X,JY], -R(X,JY)V) - ([JX,JY], -R(JX,JY)V)$$
$$= \Big(0, R(X,Y)V + JR(JX,Y)V$$
$$+ JR(X,JY)V - R(JX,JY)V\Big)$$
$$= -(0, JR(JX,Y)V + JR(X,JY)V) = (0,0).$$

(iii) mixed fields

$$N^{\mathbb{J}}(X^h, Y^v) = [X^h, Y^v] + \mathbb{J}[\mathbb{J}X^h, Y^v] + \mathbb{J}[X^h, \mathbb{J}Y^v] - [\mathbb{J}X^h, \mathbb{J}Y^v]$$
$$= (0, \nabla_X Y) + (0, J\nabla_{JX} Y) + (0, J\nabla_X JY) - (0, \nabla_{JX} JY)$$
$$= (0, \nabla_X Y + J\nabla_{JX} Y - \nabla_X Y - J\nabla_{JX} Y) = (0,0).$$

where we have used the fact that $J$ is parallel and the property $R(JX, JY) = R(X,Y)$ (see Exercise 2 of this chapter). $\qquad\square$

**Corollary 5.** *The triple* $(\mathbb{J}, \mathbb{G}, \omega)$ *defines a pseudo-Kähler structure on* $T\mathcal{M}$. *In particular* $\mathbb{J}$ *is parallel for the Levi-Civita connection.*

The following lemma describes the Levi-Civita connection $D$ of $\mathbb{G}$ in terms of the direct decomposition of $TT\mathcal{M}$.

**Lemma 15.** *Let $X$ and $Y$ two vector fields on $T\mathcal{M}$ and assume that $Y$ is projectable, then at the point $(x, V)$ we have*

$$D_X Y = (\nabla_{\Pi X} \Pi Y, \nabla_{\Pi X} KY + W(X,Y)),$$

*where*

$$W(X,Y) := \frac{1}{2}\Big(R(\Pi X, \Pi Y)V - JR(V, J\Pi X)\Pi Y - JR(V, J\Pi Y)\Pi X\Big).$$

*Proof.* We use Lemma 11 together with the Koszul formula (Lemma 1 of Chapter 1):

$$2\mathbb{G}(D_X Y, Z) = X\mathbb{G}(Y,Z) + Y\mathbb{G}(X,Z) - Z\mathbb{G}(X,Y) + \mathbb{G}([X,Y], Z)$$
$$- \mathbb{G}([X,Z], Y) - \mathbb{G}([Y,Z], X),$$

where $X$, $Y$ and $Z$ are three vector fields on $T\mathcal{M}$. From the fact that $[X^v, Y^v]$ and $\mathbb{G}(X^v, Y^v)$ vanish we have:

$$2\mathbb{G}(D_{X^v} Y^v, Z^v) = X^v\mathbb{G}(Y^v, Z^v) + Y^v\mathbb{G}(X^v, Z^v) - Z^v\mathbb{G}(X^v, Y^v)$$
$$+ \mathbb{G}([X^v, Y^v], Z^v) - \mathbb{G}([X^v, Z^v], Y^v) - \mathbb{G}([Y^v, Z^v], X^v)$$
$$= 0.$$

Moreover, taking into account that $\mathbb{G}(Y^v, Z^h)$ and similar quantities are constant on the fibres, we obtain

$$
\begin{aligned}
2\mathbb{G}(D_{X^v}Y^v, Z^h) &= X^v\mathbb{G}(Y^v, Z^h) + Y^v\mathbb{G}(X^v, Z^h) - Z^h\mathbb{G}(X^v, Y^v) \\
&\quad + \mathbb{G}([X^v, Y^v], Z^h) - \mathbb{G}([X^v, Z^h], Y^v) - \mathbb{G}([Y^v, Z^h], X^v) \\
&= -\mathbb{G}(-(\nabla_Z X)^v, Y^v) - \mathbb{G}(-(\nabla_Z Y)^v, X^v) \\
&= 0.
\end{aligned}
$$

From these last two equations we deduce that $D_{X^v}Y^v$ vanishes. Analogous computations show that $D_{X^v}Y^h$ vanishes as well. From Lemma 11 and the formula $[X, Y] = D_X Y - D_Y X$, we deduce that

$$
D_{X^h}Y^v \simeq (0, \nabla_X Y). \tag{4.13}
$$

Finally, introducing

$$
JW := \frac{1}{2}\Big( -JR(X, Y)V - R(V, JY)X - R(V, JX)Y \Big),
$$

we compute that

$$
\mathbb{G}(D_{X^h}Y^h, Z^h) = g(JW, Z)
$$

and

$$
\mathbb{G}(D_{X^h}Y^h, Z^v) = g(JZ, \nabla_X Y),
$$

from which we deduce that

$$
D_{X^h}Y^h = (\nabla_X Y, W). \tag{4.14}
$$

The conclusion of the proof follows from (4.13) and (4.14). $\qquad\square$

### 4.4.5 *Examples*

The simplest example where we may apply the construction above is where $(\mathcal{M}, J, g, \varpi)$ is the plane $\mathbb{R}^2$ equipped the flat, positive metric $\langle ., .\rangle_0$ and the complex structure $J$ coming from the identification of $\mathbb{R}^2$ with $\mathbb{C}$. We claim that in this case the structure $(T\mathbb{R}^2, \mathbb{J}, \mathbb{G}, \omega)$ just constructed is equivalent to that of the complex pseudo-Euclidean plane $(\mathbb{C}^2, J, \langle ., .\rangle_2, \omega_1)$ equipped with the structures described in Section 4.1. To see this, recalling the notation $z_1 = x_1 + iy_1$ and $z_2 = x_2 + iy_2$, consider the following linear change of coordinates

$$
q_1 := \frac{\sqrt{2}}{2}(y_1 + y_2), \qquad q_2 := \frac{\sqrt{2}}{2}(x_1 + x_2),
$$

$$p_1 := \frac{\sqrt{2}}{2}(x_1 - x_2), \qquad p_2 := \frac{\sqrt{2}}{2}(y_2 - y_1).$$

The inverse formulae are:

$$x_1 = \frac{\sqrt{2}}{2}(q_2 + p_1), \qquad y_1 = \frac{\sqrt{2}}{2}(q_1 - p_2),$$

$$x_2 = \frac{\sqrt{2}}{2}(q_2 - p_1), \qquad y_2 = \frac{\sqrt{2}}{2}(q_1 + p_2).$$

This identifies $\mathbb{C}^2$ with $T^*\mathbb{R}^2$ as described in Section 4.4.1 (recall that $(q_1, q_2)$ are coordinates on $\mathbb{R}^2$ and $dq_1 = p_1$ and $dq_2 = p_2$). Then we can check that the symplectic structure takes the following form:

$$\omega_1 = -dy_1 \wedge dx_1 + dy_2 \wedge dx_2 = dp_1 \wedge dq_1 + dp_2 \wedge dq_2 = \omega,$$

where $\omega$ is the canonical symplectic form of $T^*\mathbb{R}^2$. Using the canonical Euclidean metric of $\mathbb{R}^2$, we identify $T^*\mathbb{R}^2$ with $T\mathbb{R}^2 \simeq \mathbb{R}^2 \oplus \mathbb{R}^2$. Then one can check that the complex structure $J$ of $\mathbb{C}^2$ writes $J = j \oplus j$ according to this decomposition. It follows that the metrics $\langle .,. \rangle_2 = -|dz_1|^2 + |dz_2|^2 = \omega_1(J.,.)$ and $\mathbb{G} := \omega(\mathbb{J}.,.)$ are the same.

The next two simplest examples of Riemannian surfaces are the two Riemannian surface forms: the sphere $\mathbb{S}^2 := \mathbb{Q}^2_{0,1}$ and the hyperbolic $\mathbb{H}^2 := \mathbb{Q}^2_{2,1}$ (see Chapter 3). Their tangent bundles enjoy a beautiful geometric interpretation: the tangent bundle $T\mathbb{S}^2$ is canonically identified with the set of oriented lines of Euclidean three-space:

$$\mathbb{L}^3_0 \ni \{V + tx \mid t \in \mathbb{R}\} \simeq (x, V - \langle V, x \rangle_0 x) \in T\mathbb{S}^2.$$

Analogously, the tangent bundle $T\mathbb{H}^2$ is canonically identified with the set of oriented negative lines of three-space endowed with the metric $\langle .,. \rangle_2$:

$$\mathbb{L}^3_{2,-} \ni \{V + tx \mid t \in \mathbb{R}\} \simeq (x, V - \langle V, x \rangle_2 x) \in T\mathbb{H}^2.$$

In the next chapter we shall describe some minimal surfaces in the case where $(\mathcal{M}, g, j)$ is an oriented, Riemannian surface.

## 4.5   Exercises

(1) Let $(\mathcal{M}, J, g, \omega)$ be a pseudo-Kähler manifold and $(q, 2n - q)$ its signature. Prove that $q$ must be even. In particular, there is no pseudo-Kähler manifold of dimension two with indefinite metric.

(2) Let $(\mathcal{M}, J, g, \omega)$ a pseudo-Kähler manifold and $R$ the curvature tensor of $g$. Prove that the following identity holds: $R(JX, JY) = R(X, Y)$. (Hint: prove that $g(R(X, Y)Z, W) = g(R(X, Y)JZ, JW)$ and then use the formula $g(R(X, Y)Z, W) = g(R(Z, W)X, Y)$.)

(3) Prove that $\mathbb{CP}_p^n$ does not have constant sectional curvature. The *holomorphic* curvature of a pseudo-Kähler manifold is defined to be the sectional curvature of a complex plane, i.e. the span of $X$ and $JX$, where $X$ is a tangent vector. Given a tangent vector $X$ to $\mathbb{CP}_p^n$, prove that

$$K\big(Span(X, JX)\big) = \frac{\mathbb{G}(R(X, \mathbb{J}X)X, \mathbb{J}X)}{\mathbb{G}(X, X)g(\mathbb{J}X, \mathbb{J}X)} = 1,$$

i.e. $\mathbb{CP}_p^n$ has constant holomorphic curvature 1.

(4) For $p \neq 0$, the complex space form $\mathbb{CP}_p^n$ is an open subset of the complex projective space $\mathbb{CP}^n$. Denoting by $\iota$ the inclusion map $\mathbb{CP}_p^n \to \mathbb{CP}^n$, prove that the operator $J_\iota$ defined by the formula $d\iota \circ J_\iota := \mathbb{J} \circ d\iota$ defines a complex structure in $\mathbb{CP}_{p,\epsilon}^n$. Prove that this coincides with the complex structure constructed in this chapter, i.e. $J_\iota = \mathbb{J}$. (Hint: introduce $\tilde{\iota} : \mathbb{Q}_{2p,1}^{2n+1} \subset \mathbb{S}^{2n+1}$ defined by $\tilde{\iota}(z) = |z|_0^{-2}.z$, check that $\iota \circ \pi = \pi \circ \tilde{\iota}$ and that $d\iota \circ J = J \circ d\iota$.)

(5) Prove that $(\mathbb{CP}_p^n, g)$ is an Einstein manifold (see Exercise 1 of Chapter 3);

(6) The *scalar curvature* of a pseudo-Riemannian manifold is the trace of its Ricci curvature (see Exercise 1 of Chapter 3 for the definition). Show that if $(\Sigma, g)$ is a Riemannian surface, then $(T\Sigma, \mathbb{J}, \mathbb{G}, \omega)$ is scalar flat, i.e.

$$\operatorname{tr} Ric^{\mathbb{G}} := \sum_{k,l=1}^{4} \epsilon_l \epsilon_k \mathbb{G}(R^{\mathbb{G}}(e_l, e_k)e_k, e_l) = 0,$$

where $(e_1, e_2, e_3, e_4)$ denotes as usual an orthonormal basis of $(T\Sigma, \mathbb{G})$.

# Chapter 5

# Complex and Lagrangian submanifolds in pseudo-Kähler manifolds

In this chapter we shall describe several families of minimal submanifolds in pseudo-Kähler manifolds. Firstly, given a submanifold $\mathcal{S}$ of a pseudo-Kähler manifold $(\mathcal{M}, J, g, \omega)$, we shall say that $\mathcal{S}$ is *complex* if at any point $x$ of $\mathcal{S}$, the complex structure $J$ maps $T_x\mathcal{S}$ on itself. On the other hand, we shall say that a submanifold is *Lagrangian* if it has half the dimension of $\mathcal{M}$ and if the symplectic form $\omega$ vanishes on $T_x\mathcal{S}$, for all point $x$ of $\mathcal{S}$. By the formula $\omega = g(J., .)$, it implies that if $T_x\mathcal{S}$ is non-degenerate, $J$ maps the latter onto the normal space $N_x\mathcal{S}$. Hence, roughly speaking, complex and Lagrangian submanifolds have the two possible extreme behaviours with respect to the complex structure $J$. We shall show in the next sections that a complex submanifold is always minimal and that the extrinsic curvature of a Lagrangian submanifold takes a particular, symmetric form. Then we will refine the analysis in certain special pseudo-Kähler manifolds introduced in the previous chapter. In particular we characterize minimal, equivariant, Lagrangian submanifolds in complex pseudo-Euclidean space and complex space forms, and we give a classification of minimal Lagrangian surfaces in the tangent bundle of a Riemannian surface.

## 5.1 Complex submanifolds

**Definition 9.** Let $(\mathcal{M}, J)$ be a complex manifold. A submanifold $\mathcal{S}$ of $\mathcal{M}$ is said to be *complex* if it is stable with respect to the complex structure $J$, i.e. if at any point $x$ of $\mathcal{S}$ we have $JT_x\mathcal{S} \subset T_x\mathcal{S}$. Since $J$ is one-to-one, we have actually $JT_x\mathcal{S} = T_x\mathcal{S}$, and a complex submanifold must have even dimension.

**Theorem 22.** *Let $(\mathcal{M}, J, g, \omega)$ be a pseudo-Kähler manifold. A non-degenerate, complex submanifold of $\mathcal{M}$ is minimal.*

*Proof.* Given a tangent vector field $X$, $JX$ is also tangent to $S$. Moreover, the complex structure $J$ is parallel with respect to the flat connection $D$, so that $D_Y JX = JD_Y X$. On the other hand, since the tangent spaces are $J$-invariant, so are the normal spaces. It follows that

$$(D_Y JX)^\perp = (JD_Y X)^\perp = J(D_Y X)^\perp,$$

i.e. $h(JX, Y) = Jh(X, Y)$. Next, an easy modification of the famous Gram-Schmidt process shows that there exists an orthonormal frame $(e_1, ..., e_{2k})$ on $S$ such that $e_{2i} = Je_{2i-1}, \forall i, 1 \le i \le k$ (which proves in particular that the dimension of $S$ is even). It follows that $\epsilon_{2i} = \epsilon_{2i-1}$, and we deduce:

$$2k\vec{H} = \sum_{i=1}^{2k} \epsilon_i h(e_i, e_i) = \sum_{i=1}^{k} \epsilon_{2i-1} h(e_{2i-1}, e_{2i-1}) + \epsilon_{2i} h(e_{2i}, e_{2i})$$

$$= \sum_{i=1}^{k} \epsilon_{2i-1} \Big( h(e_{2i-1}, e_{2i-1}) + h(Je_{2i-1}, Je_{2i-1}) \Big)$$

$$= \sum_{i=1}^{k} \epsilon_{2i-1} \Big( h(e_{2i-1}, e_{2i-1}) + J^2 h(e_{2i-1}, e_{2i-1}) \Big) = 0. \qquad \square$$

It is very easy to find examples of complex submanifolds. It suffices to use the two ways used to construct submanifolds, namely immersions and submersions (see Chapter 1, Section 2), and to consider holomorphic functions rather than mere smooth functions. Then the Cauchy-Riemann equations (see Chapter 2, Section 2.4.1) say exactly that the tangent space of the submanifold is invariant by $J$.

**Proposition 11.** *Let $k < n$ be two integers, $U$ an open subset of $\mathbb{C}^k$ and $f = (f^1, ..., f^n) : U \to \mathbb{C}^n$ an immersion such that $f^j, 1 \le j \le n$, is holomorphic. Then $S := f(U)$ is a complex submanifold of $\mathbb{C}^n$ of dimension $2k$. Let $V$ be an open subset of $\mathbb{C}^n$, and $g = (g^1, ..., g^{n-k}) : V \to \mathbb{C}^{n-k}$ a submersion such that $g^j, 1 \le j \le n - k$, is holomorphic. Assume that $c \in \mathbb{C}^{n-k}$ is a regular value of $g$, i.e. the level set $S := g^{-1}(\{c\})$ is a submanifold. Then $S$ is a complex submanifold.*

The proof is elementary and left to the reader who may also check that the proposition holds true replacing the complex pseudo-Euclidean space and its open subset by complex manifolds (this last point requires to define the concept of holomorphic maps between complex manifolds, see for example [Moroianu (2007)]).

## 5.2 Lagrangian submanifolds

**Definition 10.** Let $(\mathcal{M}, \omega)$ be a symplectic manifold of dimension $2n$. The *symplectic orthogonal* of a subspace $P$ of $T_x\mathcal{M}$, where $x \in \mathcal{M}$, is the linear subspace

$$P^\omega := \{X \in T_x\mathcal{M} \mid \omega(X, Y) = 0, \forall Y \in P\}.$$

A subspace $P$ of $T_x\mathcal{M}$ is said to be *Lagrangian* if $P = P^\omega$. By the assumption of non-degeneracy of $\omega$, it follows that $\dim P = n$. A submanifold $\mathcal{L}$ of $\mathcal{M}$ is said to be *Lagrangian* if all its tangent subspaces are Lagrangian. Equivalently, $\mathcal{L}$ is Lagrangian if it has dimension $n$ and $\omega(X, Y)$ vanishes for any two vectors $X$ and $Y$ tangent to $\mathcal{L}$.

The next lemma gives two important properties of the geometry of Lagrangian submanifolds in the pseudo-Kähler setting:

**Lemma 16.** *Let $\mathcal{L}$ be a non-degenerate, Lagrangian submanifold of a pseudo-Kähler manifold $(\mathcal{M}, J, g, \omega)$ with signature $(2p, 2n - 2p)$. Then*

(i) *the signature of the induced metric of $\mathcal{L}$ is $(p, n - p)$;*
(ii) *the map*

$$T(X, Y, Z) := g(h(X, Y), JZ)$$

*is tensorial and tri-symmetric, i.e.*

$$T(X, Y, Z) = T(Y, X, Z) = T(X, Z, Y).$$

*Proof.* To prove the first point, we first claim that the Lagrangian assumption implies that $JT_x\mathcal{L} = (T_x\mathcal{L})^\perp$. To see this, observe that given $JX \in JT_x\mathcal{L}$, we have

$$\omega(X, Y) = \langle JX, Y \rangle = 0, \quad \forall Y \in T_x\mathcal{L}.$$

Therefore $JT_x\mathcal{L}$ is contained in $(T_x\mathcal{L})^\perp$, and since both spaces have dimension $n$, they are equal. Assume now that $T_x\mathcal{L}$ is non-degenerate and let $(e_1, ..., e_n)$ be an orthonormal basis of $T_x\mathcal{L}$. Therefore $(e_1, ..., e_n, Je_1, ..., Je_n)$ is an orthonormal basis of $T_x\mathcal{M}$, and denoting by $(p, n - p)$ the signature of the induced metric on $\mathcal{L}$, the signature of $g$ on $\mathcal{M}$ is of course $(2p, 2n - 2p)$.

We now study the extrinsic geometry of $\mathcal{L}$: the tensoriality of $T$ and its symmetry with respect to the first two slots follow from Propositions 1 and 2 of Chapter 1. It remains to prove for example that

$T(X, Y, Z) = T(X, Z, Y)$. From the Lagrangian assumption we have $\omega(Y, Z) = g(JY, Z) = 0$. Differentiating in the $X$ direction gives

$$0 = X(g(JY, Z)) = g(D_X JY, Z) + g(JY, D_X Z)$$
$$= g(JD_X Y, Z) + g(JY, D_X Z)$$
$$= -g(D_X Y, JZ) + g(JY, h(X, Z))$$
$$= -g(h(X, Y), JZ) + g(JY, h(X, Z))$$
$$= -T(X, Y, Z) + T(X, Z, Y),$$

and the proof is complete. □

## 5.3  Minimal Lagrangian surfaces in $\mathbb{C}^2$ with neutral metric

The simplest case of a pseudo-Kähler manifold with indefinite metric is that of $\mathbb{C}^2$ endowed with the Hermitian form

$$\langle\langle ., . \rangle\rangle_1 := -dz_1 d\bar{z}_1 + dz_2 d\bar{z}_2 = \langle ., . \rangle_2 - i\omega_1.$$

It turns out that we may characterize completely the minimal Lagrangian surfaces in this setting. We first prove a technical lemma concerning the geometry of the two-dimensional subspaces of $(\mathbb{C}^2, \langle\langle ., . \rangle\rangle_1)$:

**Lemma 17.** *Let $P$ be a plane of $(\mathbb{C}^2, \langle\langle ., . \rangle\rangle_1)$. Then the induced metric on $P$ is totally null (i.e. $\langle ., . \rangle_2|_P = 0$) if and only if $JP = P^{\omega_1}$.*

*Proof.* Suppose first that $P$ is totally null and let $X$ be a vector of $P$. For all vector $Y$ in $P$, we have

$$0 = \langle X, Y \rangle_2 = -\omega_1(JX, Y),$$

so $JX \in P^{\omega_1}$. Since it holds $\forall X \in P$, we deduce that $JP \subset P^{\omega_1}$, and the two-form $\omega_1$ being non-degenerate, $P^{\omega_1}$ is a two-dimensional subspace. Hence $JP = P^{\omega_1}$.

Conversely, if $JP = P^{\omega_1}$, then, for all vector $X$ in $P$, we have $|X|_2^2 = -\omega_1(JX, X) = 0$. By the polarization formula $2\langle X, Y \rangle_2 = |X+Y|_2^2 - |X|_2^2 - |Y|_2^2$, it implies that $P$ is totally null. □

**Remark 21.** This lemma proves in particular that a plane may be both complex and Lagrangian. This fact may sound strange to the reader familiar with Kähler geometry, where complex and Lagrangian planes are two distinct classes. More precisely, if a plane enjoys any two of the three properties: {totally null, Lagrangian, complex}, then the third one holds as well.

**Theorem 23.** *Let $\mathcal{L}$ be a minimal Lagrangian surface of $(\mathbb{C}^2, \langle\langle .,.\rangle\rangle_1)$. Then $\mathcal{L}$ is a product $\gamma_1 \times J\gamma_2 \subset P \oplus JP$, where $\gamma_1$ and $\gamma_2$ are two planar curves contained in a non-Lagrangian (and therefore non-complex) null plane $P$.*

*Proof.* By Lemma 16, the induced metric on $\mathcal{L}$ is Lorentzian, so we may use Theorem 11 of Chapter 2, and claim that $\mathcal{L}$ may be locally parametrized by an immersion of the form

$$f(u,v) = \gamma_1(u) + \tilde{\gamma}_2(v),$$

where $\gamma_1, \tilde{\gamma}_2$ are two null curves of $\mathbb{C}^2$ such that

$$\langle \gamma_1'(u), \tilde{\gamma}_2'(v)\rangle_1 \neq 0, \forall\, (u,v) \in I_1 \times I_2.$$

On the other hand the Lagrangian assumption is:

$$\omega_1(\gamma_1'(u), \tilde{\gamma}_2'(v)) = 0, \forall\, (u,v) \in I_1 \times I_2.$$

The proof relies on the analysis of the dimension of the two linear spaces $P_1 := Span\{\gamma_1'(u), u \in I_1\}$ and $P_2 := Span\{\tilde{\gamma}_2'(v), v \in I_2\}$. We first observe that $\dim P_1, \dim P_2 \geq 1$ and that the case $\dim P_1 = \dim P_2 = 1$ corresponds to the trivial case of $\mathcal{L}$ being planar. Since the rôles of $\gamma_1$ and $\tilde{\gamma}_2$ are symmetric, we may assume without loss of generality, and we do so, that $\dim P_1 \neq 1$.

Next, the Lagrangian assumption is equivalent to $P_2 \subset P_1^{\omega_1}$ and $P_1 \subset P_2^{\omega_1}$, so $\dim P_2 \leq \dim P_1^{\omega_1}$ and $\dim P_1 \leq \dim P_2^{\omega_1}$. By the non-degeneracy of $\omega_1$, it follows that $\dim P_1 \leq \dim P_2^{\omega_1} = 4 - \dim P_2 \leq 3$. We claim that in fact $\dim P_1 = 2$. To see this, assume by contradiction that $\dim P_1 = 3$. It follows that $\dim P_2 \leq \dim P_1^{\omega_1} = 1$, so the curve $\tilde{\gamma}_2$ is a straight line, which may be parametrized as follows: $\tilde{\gamma}_2(v) = e_0 v$, where $e_0$ is a null vector of $\mathbb{C}^2$. Then $\gamma_1'$ is contained in the intersection of the light cone $\mathbb{Q}_{2,0}^4$ with the hyperplane $\{e_0\}^{\omega_1}$. An easy computation, using the fact that $e_0$ is null, shows that $\mathbb{Q}_{2,0}^4 \cap \{e_0\}^{\omega_1} = \Pi_1 \cup \Pi_2$, where $\Pi_1$ and $\Pi_2$ are two null planes. Moreover, one of these planes, say $\Pi_2$, is contained in the metric orthogonal of $e_0$. By the non-degeneracy assumption,

$$\langle \gamma_1'(u), \tilde{\gamma}_2'(v)\rangle_2 = v\langle \gamma_1'(u), e_0\rangle_2 \neq 0,$$

we deduce that $\gamma_1' \in \Pi_1$, which implies that $\dim P_1 \leq 2$, a contradiction.

To conclude, observe that, using Lemma 17, $\tilde{\gamma}_2 \in P_2 \subset P_1^{\omega_1} = JP_1$. Hence we just need to set $P := P_1$ and $\gamma_2 := -J\tilde{\gamma}_2$, to get that $\gamma_1, \gamma_2 \subset P$, so that $\mathcal{L}$ takes the required expression. $\qquad\square$

## 5.4 Minimal Lagrangian submanifolds in $\mathbb{C}^n$

We now consider the complex Euclidean space of arbitrary dimension $n$, endowed with the pseudo-Hermitian form of arbitrary signature $(p, n - p)$ defined in the previous chapter:

$$\langle\langle ., . \rangle\rangle_p := -\sum_{j=1}^{p} dz_j d\bar{z}_j + \sum_{j=p+1}^{n} dz_j d\bar{z}_j.$$

We recall that this defines both a pseudo-Riemannian metric and a symplectic form:

$$\langle ., . \rangle_{2p} = \mathrm{Re}\,\langle\langle ., . \rangle\rangle_p \quad \text{and} \quad \omega_p = -\mathrm{Im}\,\langle\langle ., . \rangle\rangle_p.$$

We furthermore recall the definition of the holomorphic volume form:

$$\Omega := dz_1 \wedge ... \wedge dz_n.$$

This structure will be very useful to describe the geometry of Lagrangian submanifolds:

**Lemma 18.** *Let $(X_1, ..., X_n)$ spanning a Lagrangian, non-degenerate, linear subspace of $(\mathbb{C}^n, \langle\langle ., . \rangle\rangle_p)$. Then $\Omega(X_1, ..., X_n)$ does not vanish. Moreover, if $(Y_1, ..., Y_n)$ span the same linear subspace and has the same orientation, then the two complex numbers $\Omega(X_1, ..., X_n)$ and $\Omega(Y_1, ..., Y_n)$ have the same argument.*

*Proof.* First observe that the Lagrangian assumption is equivalent to the fact that the span of $(JX_1, ..., JX_n)$ is orthogonal to the span of $(X_1, ..., X_n)$. Therefore, if the span of $(X_1, ..., X_n)$ is non-degenerate, the $2n$ vectors $(X_1, ..., X_n, JX_1, ..., JX_n)$ span the whole space $\mathbb{C}^n$. Now, assume by contradiction that $\Omega(X_1, ..., X_n)$ vanishes. It follows that there exist $n$ complex numbers $\lambda_1, ..., \lambda_n$ such that $\sum_{j=1}^{n} \lambda_j X_j = 0$. The real part of this complex equation is

$$\sum_{j=1}^{n} \Big( \mathrm{Re}\,(\lambda_j) X_j - \mathrm{Im}\,(\lambda_j) J X_j \Big) = 0,$$

which says that the $2n$ vectors are $\mathbb{R}$-linearly dependent, a contradiction.

Next, if $(X_1, ..., X_n)$ and $(Y_1, ..., Y_n)$ span the same linear subspace and have the same orientation, there exist real constants $a_{jk}$, $1 \leq j, k \leq n$ such that $Y_j = \sum_{k=1}^{n} a_{jk} X_k$ and $\det[a_{jk}]_{1 \leq j,k \leq n} > 0$. Since

$$\Omega(Y_1, ..., Y_n) = \det[a_{jk}]_{1 \leq j,k \leq n} \Omega(X_1, ..., X_n),$$

we deduce that $\Omega(Y_1, ..., Y_n)$ and $\Omega(X_1, ..., X_n)$ have the same argument. $\square$

**Definition 11.** The *Lagrangian angle* $\beta$ of a non-degenerate, Lagrangian, oriented, linear subspace spanned by $n$ vectors $(X_1, ..., X_n)$ is the argument of $\Omega(X_1, ..., X_n)$. The *Lagrangian angle* $\beta(x)$ of an oriented, non-degenerate, Lagrangian submanifold $\mathcal{L}$ at the point $x$ is the Lagrangian angle of $T_x\mathcal{L}$. It is thus a map $\beta : \mathcal{L} \to \mathbb{R}/2\pi\mathbb{Z}$.

**Remark 22.** The definition of the Lagrangian angle, involving only $\Omega$, is independent of the choice of the metric or of the symplectic form. However a submanifold which is Lagrangian for $\omega_p$ is not Lagrangian *a priori* for $\omega_{p'}$, $p' \neq p$.

**Theorem 24.** *Let $\mathcal{L}$ be a non-degenerate, Lagrangian submanifold of $(\mathbb{C}^n, \langle\langle ., .\rangle\rangle_p)$ with Lagrangian angle $\beta$ and mean curvature vector $\vec{H}$. Then the following holds:*

$$n\vec{H} = J\nabla\beta,$$

*where $\nabla$ denotes the gradient operator with respect to the induced metric.*

*Proof.* Let $(e_1, ..., e_n)$ a local orthonormal frame of $\mathcal{L}$. The Lagrangian assumption implies that it is also a Hermitian frame, i.e.

$$[\langle\langle e_j, e_k\rangle\rangle_p]_{1 \leq j,k \leq n} = diag(\epsilon_1, ..., \epsilon_n).$$

In particular, given any vector $X$ of $\mathbb{C}^n$, we have

$$X = \sum_{j=1}^{p} \epsilon_j \langle\langle e_j, X\rangle\rangle_p e_j.$$

It is sufficient to prove that $\langle n\vec{H}, Je_j\rangle_{2p} = \langle J\nabla\beta, Je_j\rangle_{2p}$, i.e. $\langle n\vec{H}, Je_j\rangle_{2p} = d\beta(e_j)$. Differentiating the identity $e^{i\beta} = \Omega(e_1, ..., e_n)$ with respect to the vector $e_j$, we have

$$id\beta(e_j)e^{i\beta} = \sum_{k=1}^{n} \Omega(e_1, ..., D_{e_j}e_k, ..., e_n)$$

$$= \sum_{k=1}^{n} \Omega\left(e_1, ..., \sum_{l=1}^{n}\langle\langle e_l, D_{e_j}e_k\rangle\rangle_p e_l, ..., e_n\right)$$

$$= \sum_{k=1}^{n} \Omega(e_1, ..., \epsilon_k\langle\langle e_k, D_{e_j}e_k\rangle\rangle_p e_k, ..., e_n)$$

$$= \sum_{k=1}^{n} \epsilon_k\langle\langle e_k, D_{e_j}e_k\rangle\rangle_p e^{i\beta},$$

hence

$$id\beta(e_j) = \sum_{k=1}^{n} \epsilon_k \langle\langle e_k, D_{e_j} e_k \rangle\rangle_p.$$

Recalling that $\langle\langle .,. \rangle\rangle_p = \langle .,. \rangle_{2p} - i\omega_p = \langle .,. \rangle_{2p} - i\langle J.,. \rangle_{2p}$, we have

$$\langle\langle e_k, D_{e_j} e_k \rangle\rangle_p = \langle e_k, D_{e_j} e_k \rangle_{2p} - i\langle Je_k, D_{e_j} e_k \rangle_{2p}.$$

Differentiating the relation $\langle e_k, e_k \rangle_{2p} = \epsilon_k$ in the direction $e_j$ yields $\langle e_k, D_{e_j} e_k \rangle_{2p} = 0$, so that, taking into account that $Je_j$ is a normal vector to $\mathcal{L}$,

$$\langle\langle e_k, D_{e_j} e_k \rangle\rangle_p = i\langle Je_k, D_{e_j} e_k \rangle_{2p} = i\langle Je_k, h(e_j, e_k) \rangle_{2p}.$$

By Lemma 16, we conclude that

$$d\beta(e_j) = \sum_{k=1}^{n} \epsilon_k \langle Je_j, h(e_k, e_k) \rangle_{2p} = \langle Je_j, n\vec{H} \rangle_{2p}. \qquad \square$$

**Corollary 6.** *A Lagrangian submanifold $\mathcal{L}$ of $\mathbb{C}^n$ is minimal if and only if it has constant Lagrangian angle.*

**Remark 23.** Let $\mathcal{L}$ be a Lagrangian submanifold with Lagrangian angle $\beta$ and consider the rotation of $\mathbb{C}^n$ whose complex matrix is $e^{i\beta_0} Id$. Then it is easy to check that the image $e^{i\beta_0}\mathcal{L}$ of $\mathcal{L}$ by this rotation has Lagrangian angle $\beta + n\beta_0$. It means that there is no loss of generality to study minimal Lagrangian submanifolds with *vanishing* Lagrangian angle $\beta \equiv 0$ (the latter are called *special Lagrangian* in the Riemannian case), since we can get any other constant angle Lagrangian by a suitable rotation of the ambient space.

### 5.4.1 *Lagrangian graphs*

In this section, it will be convenient to set $\epsilon_j = -1, 1 \leq j \leq p$ and $\epsilon_j = 1, p+1 \leq j \leq n$. In particular, we can write

$$\langle\langle .,. \rangle\rangle_p = \sum_{j=1}^{n} \epsilon_j dz_j d\bar{z}_j.$$

**Theorem 25 ([Dong (2009)]).** *Let $U$ be some open subset of $\mathbb{R}^n$, $u$ a smooth, real-valued function defined on $U$ and $\mathcal{L}$ be the graph of its gradient, i.e. the image of the immersion*

$$f : U \to \mathbb{C}^n$$
$$x \mapsto x + i\nabla u = (x_1 + i\epsilon_1 u_{x_1}, ..., x_n + i\epsilon_n u_{x_n})$$

*($\nabla u$ is the gradient of $u$ with respect to the metric $\langle .,.\rangle_p$ of $\mathbb{R}^n$). Then $\mathcal{L}$ is Lagrangian. Moreover, its Lagrangian angle function is given by the formula*

$$\beta = \arg \det \left( Id + i \left[ \epsilon_j u_{x_j x_k} \right]_{1 \leq j,k \leq n} \right).$$

*Proof.* For all $j$, $1 \leq j \leq n$, we compute

$$f_{x_j} = (i\epsilon_1 u_{x_j x_1}, ..., 1 + i\epsilon_j u_{x_j x_j}, ..., i\epsilon_n u_{x_j x_n}),$$

so that

$$\omega_p(f_{x_j}, f_{x_k}) = \epsilon_j^2 u_{x_j x_k} - \epsilon_k^2 u_{x_k x_j} = 0,$$

hence the immersion $f$ is Lagrangian. Next, we compute:

$$\beta = \arg \Omega(f_{x_1}, ..., f_{x_n}) = \arg \det \left( Id + i \left[ \epsilon_j u_{x_j x_k} \right]_{1 \leq j,k \leq n} \right),$$

which completes the proof. $\qquad\square$

**Remark 24.** If $n = 2$, denoting the two possible Hermitian forms by $\epsilon_1 dz_1 d\bar{z}_1 + dz_2 d\bar{z}_2$, we have

$$\beta = \arg \left( 1 - \epsilon_1 (u_{x_1 x_1} u_{x_2 x_2} - u_{x_1 x_2}^2) + i \left( \epsilon_1 u_{x_1 x_1} + u_{x_2 x_2} \right) \right).$$

Thus the corresponding Lagrangian surface has vanishing Lagrangian angle if and only if

$$\epsilon_1 u_{x_1 x_1} + u_{x_2 x_2} = 0.$$

In other words, $u$ is harmonic for the Laplacian of the metric $\epsilon_1 dx_1^2 + dx_2^2$. In the definite case (resp. in the indefinite case), this is the Laplace equation (resp. the wave equation). Both PDEs are easy to solve locally: the general solution of the Laplace equation takes the form $u(x_1, x_2) = \operatorname{Re} F(x_1 + ix_2)$, where $F$ is a holomorphic function, and the general solution of the wave equation is $u(x_1, x_2) = F_1(x_1 + x_2) + F_2(x_1 - x_2)$, where $F_1$ and $F_2$ are two functions of the real variable. In the indefinite case, setting $u := x_1 + x_2$ and $v := x_1 - x_2$, we see that the corresponding graph is the product of the one-dimensional graphs $(u, F_1'(u))$ and $(v, F_2'(v))$, which are curves in the null planes $P := \{x_1 - x_2 = 0, y_1 + y_2 = 0\}$ and $JP = \{x_1 + x_2 = 0, y_1 - y_2 = 0\}$ respectively. We therefore recover a special case of Theorem 23. We also recover the observation made in Section 2.4.3 of Chapter 2: in the positive case, the minimal surfaces are analytic, while in the indefinite case, they only need to be twice differentiable. This phenomenon still occurs in higher dimension, since the PDE $\arg \Omega(f_{x_1}, ..., f_{x_n}) = \arg \det \left( Id + i \left[ \epsilon_j u_{x_j x_k} \right]_{1 \leq j,k \leq n} \right) = 0$ turns out to be elliptic if $p = 0$ and hyperbolic otherwise (see [Dong (2009)] for more details).

### 5.4.2    *Equivariant Lagrangian submanifolds*

The group $SO(p, n-p)$ acts in a canonical way on the space $(\mathbb{C}^n, \langle\langle ., . \rangle\rangle_p)$ as follows: for $z = x + iy \in \mathbb{C}^n$ and $M \in SO(p, n-p)$ we simply set $Mz := Mx + iMy$. Of course we have $\langle\langle Mz, Mz' \rangle\rangle_p = \langle\langle z, z' \rangle\rangle_p$, so $SO(p, n-p)$ can be identified with a subgroup of $U(p, n-p)$. This subgroup is of interest for us because its orbits are $(n-1)$-dimensional submanifolds (except possibly at the origin), while Lagrangian submanifolds have dimension $n$. Therefore we may hope to be able to describe Lagrangian submanifolds which are $SO(p, n-p)$-invariant. We recall the observation done in Chapter 3: the orbits of the action $SO(p, n-p)$ on $\mathbb{R}^n$ take the form:

$$\mathbb{Q}_{p,c}^{n-1} := \left\{ x \in \mathbb{R}^n \mid \langle x, x \rangle_p = c \right\}.$$

**Theorem 26.** *Let $\mathcal{L}$ be a $SO(p, n-p)$-equivariant Lagrangian submanifold of $\mathbb{C}^n$. Then it is locally the image of an immersion of the form*

$$
\begin{aligned}
f : I \times \mathbb{Q}_{p,\epsilon}^{n-1} &\to \quad \mathbb{C}^n \\
(s, x) &\mapsto \gamma(s) x,
\end{aligned}
$$

*where $\epsilon = 1$ or $-1$ and $\gamma : I \to \mathbb{C}^*$ is a planar curve. Moreover, the Lagrangian angle of $\mathcal{L}$ is given by*

$$\beta = \arg(\gamma' \gamma^{n-1}).$$

**Remark 25.** In the definite, two-dimensional case ($p = 0, n = 2$), the $SO(2)$-action mentioned in the theorem above is not the only possible one, and there do exist Lagrangian surfaces of $\mathbb{C}^2$ equivariant by *another $SO(2)$* action. For example, let $\gamma(s) = (\gamma_1(s), \gamma_2(s))$ be a regular curve of the sphere $\mathbb{S}^3$. Then the map $f(s, t) = (\gamma_1(s)e^{it}, \gamma_2(s)e^{it})$ is an immersion as soon as $\langle \gamma', J\gamma \rangle$ does not vanish and is Lagrangian. Such an immersion is equivariant by the action $M(z_1, z_2) = (Mz_1, Mz_2), M \in SO(2)$. These surfaces have been studied in [Pinkall (1985)], where they are called *Hopf surfaces.*

*Proof of Theorem 26.*
     **First case:** $n = 2$.
     Recall that the metric of $\mathbb{C}^2$ is $\langle\langle ., . \rangle\rangle_p = \epsilon_1 dz_1 d\bar{z}_1 + dz_2 d\bar{z}_2$ with $\epsilon_1 = 1$ or $-1$. Introducing $M_{\epsilon_1} := \begin{pmatrix} 0 & -\epsilon_1 \\ 1 & 0 \end{pmatrix}$, we have

$$SO(2) = \{e^{M_1 t}, t \in \mathbb{R}\} \quad \text{and} \quad SO(1,1) = \{e^{M_{-1} t}, t \in \mathbb{R}\}.$$

A surface of $\mathbb{C}^2$ which is $SO(2)$ or $SO(1,1)$-equivariant may be locally parametrized by an immersion of the form

$$f(s, t) = e^{M_{\epsilon_1} t}(z_1(s), z_2(s)).$$

We first compute the first derivatives of the immersion:
$$f_s = e^{M_{\epsilon_1}t}(z_1', z_2'),$$
$$f_t = e^{M_{\epsilon_1}t}M_{\epsilon_1}(z_1, z_2) = e^{M_{\epsilon_1}t}(-\epsilon_1 z_2, z_1).$$

Therefore the Lagrangian condition yields:
$$\begin{aligned} 0 = \omega_p(f_s, f_t) &= \omega_p((z_1', z_2'), (-\epsilon z_2, z_1)) \\ &= -\text{Im}\,(z_1' \bar{z}_2) + \text{Im}\,(z_2' \bar{z}_1) \\ &= \frac{d}{ds}\text{Im}\,(z_2 \bar{z}_1). \end{aligned}$$

Hence $z_1 \bar{z}_2$ must be constant. Observe that there is no loss of generality in assuming that $\text{Im}\,(z_1 \bar{z}_2)$ vanishes: otherwise, it suffices to replace $f$ by $\tilde{f} := \begin{pmatrix} 1 & 0 \\ 0 & e^{i\arg z_2(0)} \end{pmatrix} f$, which has the same geometry. Thus, $z_1$ and $z_2$ have the same argument. Next introduce polar coordinates $z_1 = r_1 e^{i\varphi}$ and $z_2 = r_2 e^{i\varphi}$ and consider separately the definite and indefinite cases:

*The definite case $p = 0$:*
The second coordinate of $f$ is
$$z_2(s)\cos t + z_1(s)\sin t = (r_2(s)\cos t + r_1(s)\sin t)e^{i\varphi(s)}.$$

Clearly, $\forall s \in I$, there exists $t(s) \in \mathbb{R}$ such that $r_2(s)\sin t(s) + r_1(s)\cos t(s) = 0$, hence the second coordinate of $f$ vanishes at $(s, t(s))$. Setting $\gamma(s) := z_1(s)\cos t(s) - z_2(s)\sin t(s)$, i.e. $\gamma(s)$ is the first coordinate of $f$ at $(s, t(s))$, we see that $f(s, t) = e^{M_1(t-s(t))}(\gamma(s), 0)$. Hence the immersion $\tilde{f}(s, t) := e^{M_1 t}(\gamma(s), 0) = (\gamma\cos t, \gamma\sin t)$ parametrizes the same surface as $f$, and we get the required parametrization for the surface $\mathcal{L}$.

*The indefinite case $p = 1$:*
We first observe that $r_1 \neq r_2$ since otherwise the immersion would be degenerate. If $r_1 > r_2$, there exists $t(s)$ such that $r_2(s)\cosh t(s) + r_1(s)\sinh t(s) = 0$, hence the second coordinate of $f$ vanishes at $(s, t(s))$. Analogously to the definite case, we set $\gamma(s) = r_1(s)\cosh t(s) + r_2(s)\sinh t(s)$, and as before, we check that $\tilde{f}(s, t) := (\gamma(s)\cosh t, \gamma(s)\sinh t)$ parametrizes the same surface as $f$. The argument is strictly analogous if $r_1 < r_2$: we find $t(s)$ in order to make the first coordinate vanish and find $\tilde{f}(s, t) = (\gamma(s)\sinh t, \gamma(s)\cosh t)$.

**Second case: $n \geq 3$.** First, set three different indexes $j, k$ and $l$ (here we are using the assumption $n \geq 3$!) and consider the two matrices $M_{jl}$ and $M_{kl}$ defined by
$$M_{jl}e_j = e_l, \qquad M_{jl}e_l = \epsilon_j \epsilon_l e_j \quad \text{and} \quad M_{jl}e_m = 0 \text{ for } m \neq j, l,$$

and

$$M_{kl}e_k = e_l, \qquad M_{kl}e_l = \epsilon_k \epsilon_l e_k \quad \text{and} \quad M_{kl}e_m = 0 \text{ for } m \neq k, l.$$

The reader may check that $M_{jl}$ and $M_{kl}$ are skew with respect to $\langle ., . \rangle_p$. Hence, given a point $z$ in $\mathcal{L}$, the two curves $s \mapsto e^{M_{jl}s}$ and $s \mapsto e^{M_{kl}s}$ belong to $SO(p, n-p)$. By the equivariance assumption, it follows that the curves $s \mapsto e^{M_{jl}s}z$ and $s \mapsto e^{M_{kl}s}z$ belong to $\mathcal{L}$, so the two vectors $M_{jl}z$ and $M_{kl}z$ are tangent to $\mathcal{L}$ at $z$. Moreover the Lagrangian assumption yields

$$0 = \omega_p(M_{jl}z, M_{kl}z) = \operatorname{Re} z_j \operatorname{Im} z_k - \operatorname{Re} z_k \operatorname{Im} z_j.$$

Since this holds for any pair of indexes $(j, k)$, it follows that $\operatorname{Re} z$ and $\operatorname{Im} z$ are collinear. Therefore there exist $\varphi \in \mathbb{R}$ and $y \in \mathbb{R}^n$ such that $z = e^{i\varphi}y$. Let $r > 0$ and $x \in \mathbb{Q}_{p,\epsilon}^{n-1}$ such that $y = rx$, and set $\gamma := re^{i\varphi}$. By the equivariance assumption, the $(n-1)$-dimensional quadric $\gamma \mathbb{Q}_{p,\epsilon}^{n-1}$ of $\mathbb{C}^n$ is contained in $\mathcal{L}$. Finally, since $\mathcal{L}$ is $n$-dimensional, it must be locally foliated by a one-parameter family of quadrics $\gamma(s)\mathbb{Q}_{p,\epsilon}^{n-1}$, which proves the first part of the theorem (characterization of equivariant Lagrangian submanifolds).

It remains to calculate the Lagrangian angle: let $f$ be an immersion as described in the statement of the theorem, $x$ a point of $\mathbb{Q}_{p,\epsilon}^{n-1}$ and $(e_1, ..., e_{n-1})$ an oriented orthonormal basis of $T_x\mathbb{Q}_{p,\epsilon}^{n-1}$. Setting

$$X_j := \gamma e_j \quad \text{and} \quad X_n := \gamma' x,$$

it is easy to check that $(X_1, ..., X_n)$ is a basis of $T_{\gamma x}\mathcal{L}$. Then, we calculate

$$\omega_p(X_j, X_k) = \langle JX_j, X_k \rangle_{2p} = \langle i\gamma, \gamma \rangle \langle e_j, e_k \rangle_p = 0,$$

$$\omega_p(X_j, X_n) = \langle JX_j, X_n \rangle_{2p} = \langle i\gamma, \gamma' \rangle \langle e_j, x \rangle_p = 0,$$

which shows that $\omega_p$ vanishes on $T_{\gamma x}\mathcal{L}$, i.e. $\mathcal{L}$ is Lagrangian. Finally, we get the Lagrangian angle of $\mathcal{L}$ as follows:

$$\begin{aligned} e^{i\beta} &= \Omega(X_1, ..., X_n) \\ &= \Omega(\gamma e_1, ..., \gamma e_{n-1}, \gamma' x) \\ &= \gamma' \gamma^{n-1} \Omega(e_1, ..., e_{n-1}, x) = \gamma' \gamma^{n-1}. \qquad \square \end{aligned}$$

From Theorem 26, it is straightforward to describe equivariant minimal Lagrangian submanifolds: $\beta$ vanishes if and only if $\operatorname{Im} \gamma' \gamma^{n-1} = 0$, which we easily integrate to get $\operatorname{Im} \gamma^n = c$ for some real constant $c$. If $c$ vanishes, the curve $\gamma$ is made up of $n$ straight lines passing through the origin, and the corresponding Lagrangian submanifold is nothing but the union of $n$ linear spaces of $\mathbb{C}^n$. If $c$ does not vanish, the curve is a made up of $2n$

pieces, each of one contained in an angular sector $\{\varphi_0 < \arg \gamma < \varphi_0 + \frac{\pi}{n}\}$. If $n = 2$, they are hyperbolae. Summing up, we have obtained the following characterization of equivariant, minimal Lagrangian submanifolds:

**Corollary 7.** *Let $\mathcal{L}$ be a connected, minimal Lagrangian submanifold of $(\mathbb{C}^n, \langle\langle .,. \rangle\rangle_p)$ which is $SO(p, n-p)$-equivariant. Then $\mathcal{L}$ is an open subset of one of the following submanifolds:*

(i) *an affine Lagrangian $n$-plane;*
(ii) *the* Lagrangian catenoid

$$\{\gamma.x \in \mathbb{C}^n \mid x \in \mathbb{Q}_{p,\epsilon}^{n-1}, \gamma \in \mathbb{C}, \operatorname{Im} \gamma^n = c\},$$

*where $c$ is a non-vanishing real constant.*

The Lagrangian catenoid has been described for the first time in [Harvey, Lawson (1982)] in the positive case $p = 0$, although this name was introduced in [Castro, Urbano (1999)], where it was also proved that it is the only minimal, Lagrangian submanifold which is foliated by $(n-1)$-dimensional spheres (cf also [Anciaux, Castro, Romon (2006)]).

### 5.4.3   *Lagrangian submanifolds from evolving quadrics*

This section describes a class of Lagrangian submanifolds which generalize the former ones and follows ideas from [Joyce (2001)] (see also [Lee, Wang (2009)],[Joyce, Lee, Tsui (2010)]). Consider a real, invertible $n \times n$ matrix $M$ which is self-adjoint with respect to $\langle .,. \rangle_p$, i.e. $\langle Mx, y \rangle_p = \langle x, My \rangle_p, \forall x, y \in \mathbb{R}^n$. The one-parameter family of complex $n \times n$ matrices $e^{iMs}$, $s \in \mathbb{R}$ is defined by the two following equivalent properties:

(i)

$$e^{iMs} = \sum_0^\infty \frac{(iM)^k}{k!} s^k;$$

(ii) the curve $s \mapsto e^{iMs}$ is the solution of the ODE

$$\frac{d}{ds}(e^{iMs}) = iMe^{iMs} = e^{iMs}iM$$

with initial condition $e^{iM0} = Id$.

**Remark 26.** If there exists an orthonormal basis $(e_1, ..., e_n)$ of eigenvectors for $M$, the matrix $e^{iMs}$ expressed in this basis is simply $diag(e^{i\lambda_1 s}, ..., e^{i\lambda_n s})$. However, if the metric $\langle .,. \rangle_p$ is indefinite, there does not necessarily exist such a basis.

**Lemma 19.** $e^{iMs}$ *belongs to* $U(p, n-p)$, *i.e. it preserves the Hermitian product* $\langle\langle ., . \rangle\rangle_p$ :

$$\langle\langle e^{iMs}z, e^{iMs}z\rangle\rangle_p = \langle\langle z, z'\rangle\rangle_p, \quad \forall z, z' \in \mathbb{C}^n. \tag{5.1}$$

*Moreover,*

$$\det_{\mathbb{C}}[e^{iMs}] = e^{i\,\mathrm{tr}Ms}. \tag{5.2}$$

*Proof.* Observe first that Formula (5.1) holds true for $s = 0$; hence, the right hand side term of (5.1) being independent of $s$, it is sufficient to prove the left hand side is constant with respect to $s$. Differentiating it, we get

$$\langle\langle iMe^{iMs}z, e^{iMs}z'\rangle\rangle_p + \langle\langle e^{iMs}z, iMe^{iMs}z'\rangle\rangle_p.$$

Writing $e^{iMs}z = A + iB$ and $e^{iMs}z' = A' + iB'$, with $A, B, A', B' \in \mathbb{R}^n$, we have, using the properties of the Hermitian product,

$$\langle\langle iM(A + iB), A' + iB'\rangle\rangle_p + \langle\langle A + iB, iM(A' + iB')\rangle\rangle_p$$

$$= -\langle MB, A'\rangle_p + \langle MA, B'\rangle_p - \langle A, MB'\rangle_p + \langle B, MA'\rangle_p$$

$$+ i\left(\langle MA, A'\rangle_p + \langle MB, B'\rangle_p - \langle A, MA'\rangle_p - \langle B, MB'\rangle_p\right)$$

which vanishes since $M$ is self-adjoint.

The second assertion of the lemma is proved in a similar manner, using Lemma 3 of Chapter 1: writing $u(s) := \det_{\mathbb{C}}[e^{iMs}]$, we use the fact that $\frac{d}{dt}(e^{iM(s+t)})\big|_{t=0} = e^{iMs}iM$ to find that:

$$\frac{du(s)}{ds} = \frac{d}{dt}u(s+t)\bigg|_{t=0} = u(s)\mathrm{tr}(e^{-iMs}e^{iMs}iM) = u(s)i\mathrm{tr}M.$$

Hence $u$ is solution of the equation $\frac{du}{ds} = ui\,\mathrm{tr}M$ with initial condition $u(0) = 1$, that is

$$\det_{\mathbb{C}}[e^{iMs}] = u(s) = e^{i\,\mathrm{tr}Ms}. \qquad \square$$

**Theorem 27.** *Let* $c \in \mathbb{R}$ *such that the quadric*

$$S := \{x \in \mathbb{R}^n | \langle x, Mx\rangle_p = c\}$$

*is a non-degenerate hypersurface of* $(\mathbb{R}^n, \langle ., .\rangle_p)$ *and* $r(s)$ *a positive function on an interval* $I$ *of* $\mathbb{R}$. *Then the immersion*

$$f : I \times S \to \mathbb{C}^n$$
$$(s, x) \mapsto r(s)e^{iMs}x$$

*is a Lagrangian and its Lagrangian angle is given by*

$$\beta = \mathrm{tr}Ms + \arg\left(c\frac{r'}{r} + i|Mx|_p^2\right) + \pi/2.$$

*Proof.* Let $(e_1, ..., e_{n-1})$ be an orthonormal basis of $T_x S = (Mx)^\perp$, that we complete by $e_n$ in such a way that $(e_1, ..., e_n)$ is an oriented, orthonormal basis of $\mathbb{R}^n$. Hence $e_n$ is collinear to $Mx$ and, setting $\epsilon_n = |e_n|_p^2$ we have

$$Mx = \epsilon_n \langle Mx, e_n \rangle_p e_n.$$

Set $\mathcal{L} := f(I \times S)$. We obtain a basis of $T_z \mathcal{L}$, at a point $z = re^{iMs}x$, setting

$$Z_j := e^{iMs} e_j \quad \text{and} \quad Z_n := (r' + riM)e^{iMs}x.$$

Using the fact that $e^{iMs} \in U(p, n - p)$ (Lemma 19), we check that

$$\omega_p(Z_j, Z_n) = \omega_p(e_j, (r' + irM)x)$$

$$= r'\omega_p(e_j, x) + r\langle e_j, Mx \rangle_p = 0$$

and

$$\omega_p(Z_j, Z_k) = \omega_p(e_j, e_k) = 0,$$

hence the immersion $f$ is Lagrangian. To complete the proof, we compute

$$\Omega(Z_1, ..., Z_n) = \Omega(e^{iMs}e_1, ..., e^{iMs}e_{n-1}, (r' + irM)e^{iMs}x)$$

$$= i \det_{\mathbb{C}}[e^{iMs}] \det_{\mathbb{C}}(e_1, ..., e_{n-1}, (r' + irM)x)$$

$$= i \det_{\mathbb{C}}[e^{iMs}] \left( r'\epsilon_n \langle x, e_n \rangle_p + ir\epsilon_n \langle Mx, e_n \rangle \right).$$

Using Lemma 19 and the fact that

$$\langle x, e_n \rangle_p = \frac{\langle x, Mx \rangle_p}{\epsilon_n \langle Mx, e_n \rangle_p} = \frac{c}{\epsilon_n \langle Mx, e_n \rangle_p},$$

we get

$$\Omega(Z_1, ..., Z_n) = ie^{i\operatorname{tr}Ms} \frac{r}{\langle Mx, e_n \rangle} \left( c\frac{r'}{r} + i\epsilon_n \langle Mx, e_n \rangle^2 \right).$$

We deduce, using the fact that $\epsilon_n \langle Mx, e_n \rangle^2 = |Mx|_p^2$,

$$\beta = \arg(\Omega(Z_1, ..., Z_n))$$

$$= \frac{\pi}{2} + \operatorname{tr}Ms + \arg\left( c\frac{r'}{r} + i|Mx|_p^2 \right),$$

which is the required formula. $\qquad\square$

**Example 7.** Assume that $M = Id$ and $c = 1$. Then $f$ becomes

$$f : I \times \mathbb{Q}_{p,\epsilon}^{n-1} \to \quad \mathbb{C}^n$$
$$(s, x) \quad \mapsto r(s)e^{is}x.$$

In particular the image of the immersion is a $SO(p, n - p)$-equivariant submanifold as in Section 5.4.2.

**Corollary 8.** *The Lagrangian immersion f introduced in Theorem 27 above is minimal if and only if one of the three statements holds:*

(i) $\operatorname{tr} M = 0$ *and the function r is constant;*
(ii) $\operatorname{tr} M = 0$ *and the constant c vanishes;*
(iii) *the image of f is a part of the Lagrangian catenoid described in the previous section.*

*Proof.* The Lagrangian angle $\beta$ must be constant, so the term $\arg\left(c\frac{r'}{r} + i|Mx|_p^2\right)$ must be independent of $x$. This happens if and only either $r'$ or $c$ vanishes, or both $|Mx|_p^2$ and $\frac{r'}{r}$ are constant. If $r'$ or $c$ vanish, the first term $\operatorname{tr} Ms$ of $\beta_f$ must be constant as well, hence we must have $\operatorname{tr} M = 0$. These are the first two cases of the corollary. Suppose now $|Mx|_p^2$ is constant on $\mathcal{S}$, i.e.

$$\forall x \in \mathbb{R}^n \text{ such that } \langle Mx, x\rangle_p = c, \ |Mx|_p^2 = c'.$$

Since $M$ is invertible, it is equivalent to

$$\forall y \in \mathbb{R}^n \text{ such that } \langle y, M^{-1}y\rangle_p = c, \ |y|_p^2 = c'.$$

It follows that the quadric $\{\langle y, M^{-1}y\rangle_p = c\}$ is contained in the quadric $\mathbb{Q}_{p,c'}^{n-1}$, hence $M^{-1}$ is a multiple of the identity and so is $M$. Hence the immersion is equivariant and we are in the situation described in Example 7 above. The result then follows from Corollary 7.      □

**Example 8.** Set $n = 2$, $p = 1$ and $M = \begin{pmatrix} 0 & -1 \\ 1 & 0 \end{pmatrix}$. Since $\operatorname{tr} M = 0$ the immersion $f(s, x) = r(s)e^{iMs}x$ is minimal if $c$ vanishes or if $r$ is constant. The case of vanishing $c$ is trivial: the quadric $\mathcal{S}$ reduces to the union of the two straight lines $\{x_1 = 0\}$ and $\{x_2 = 0\}$, and the image of $f$ is the union the two complex planes $\{z_1 = 0\}$ and $\{z_2 = 0\}$.

In the case of non-vanishing $c$, we set

$$\mathcal{S} = \{x \in \mathbb{R}^2 | \langle x, Mx\rangle_1 = 2x_1x_2 = c\}$$

is an hyperbola which may be parametrized by $t \mapsto (e^t, \frac{2}{c}e^{-t})$. On the other hand

$$e^{iMs} = \begin{pmatrix} \cosh s & -i\sinh s \\ i\sinh s & \cosh s \end{pmatrix}.$$

so, setting $r = 1$, we are left with the immersion

$$f(s, t) = \left(e^t\cosh s - i\frac{2}{c}e^{-t}\sinh s, \ \frac{2}{c}e^{-t}\cosh s + ie^t\sinh s\right).$$

Observe that
$$|f_s|_2^2 = -|f_t|_2^2 = e^{2t} - \frac{4}{c^2}e^{-2t} \quad \text{and} \quad \langle f_s, f_t \rangle_2 = 0,$$
i.e. $(s, t)$ are conformal coordinates. Therefore, the coordinates $(u, v)$ defined by $u = s + t$ and $v = s - t$ are null and we get
$$f(u, v) = \frac{1}{2}\left(e^u + i\frac{2}{c}e^{-u}, \frac{2}{c}e^{-u} + ie^u\right) + \frac{1}{2}\left(e^{-v} - i\frac{2}{c}e^v, \frac{2}{c}e^v - ie^{-v}\right)$$
Therefore $f$ takes the form $f(u, v) = \gamma_1(u) + J\gamma_2(v)$ where
$$\gamma_1(u) := \frac{1}{2}\left(e^u + \frac{2}{c}ie^{-u}, e^{-u} + \frac{2}{c}ie^u\right)$$
and
$$\gamma_2(v) := \frac{-1}{2}\left(e^v + i\frac{2}{c}e^{-v}, e^v + i\frac{2}{c}e^{-v}\right)$$
are two hyperbolae in the null plane $P = \{x_1 - y_2 = 0, x_2 - y_1 = 0\}$. We therefore recover a special case of Theorem 23.

**Example 9.** In the definite case, since the metric $\langle ., . \rangle_0$ is positive, there exists an orthonormal basis of eigenvectors of $M$. So we may assume without loss of generality that $M = diag(\lambda_1, ..., \lambda_n)$, where the $\lambda_j$s are real constants. It follows that a point $z$ of $\mathcal{L}$ takes the form
$$(x_1 e^{i\lambda_1 s}, ..., x_n e^{i\lambda_n s}),$$
where $\sum_{j=1}^n \lambda_j x_j^2 = c$. We observe furthermore that $\mathcal{L}$ is a properly immersed submanifold if and only if all the coefficients $\lambda$ are rationally related. In this case we may assume without loss of generality that they are integer numbers. This case is studied in [Lee, Wang (2009)]. Observe moreover that $c$ cannot vanish (otherwise $\mathcal{S}$ reduces to the origin), so by Theorem 8, $\mathcal{L}$ is minimal if and only if $\text{tr}\, M = 0$. Example 8 above proves that the situation is richer in the indefinite case, since there may not exist an orthonormal basis of eigenvectors.

## 5.5 Minimal Lagrangian submanifols in complex space forms

In this section we study equivariant minimal Lagrangian surfaces in the complex space forms $\mathbb{CP}_p^n$ introduced in Chapter 4. The definite cases $\mathbb{CP}^n$ and $\mathbb{CH}^n$ have been studied in [Castro, Montealegre, Urbano (1999)], [Anciaux (2006)], [Castro, Li, Urbano (2006)] and [Haskins, Kapouleas (2010)]. We shall see that the situation shares many features with that of equivariant minimal hypersurfaces in space forms. In particular the problem is reduced to the study of certain curves in surface forms.

### 5.5.1    *Lagrangian and Legendrian submanifolds*

In this section, we set $\mathbb{Q} := \mathbb{Q}_{2p,1}^{2n+1}$ for sake of brevity.

**Definition 12.** A $n$-dimensional submanifold $\mathcal{L}$ of $\mathbb{Q}$ is said to be *Legendrian* if it is *integrable* with respect to the hyperplane distribution $p \mapsto (Jz)^{\perp}$, i.e.

$$T_z\mathcal{L} \subset (Jz)^{\perp}, \quad \forall z \in \mathcal{L}.$$

**Example 10.** The real space form $\bar{\mathbb{Q}}_{p,1}^{n} := \mathbb{Q} \cap \mathbb{R}^{n+1}$ is a Legendrian submanifold of $\mathbb{Q}$. More generally, if $E$ is a Lagrangian subspace of $\mathbb{C}^{n+1}$, the quadric $\mathbb{Q} \cap E$ is Legendrian.

The next lemma shows that the properties of Legendrian submanifolds of $\mathbb{Q}$ are similar to those of Lagrangian submanifolds of pseudo-Euclidean space:

**Lemma 20.** *A Legendrian submanifold $\mathcal{L}$ of $\mathbb{Q}$ is isotropic as a submanifold of $\mathbb{C}^{n+1}$, i.e. the symplectic form $\omega_p$ vanishes on $\mathcal{L}$. Moreover, if $\mathcal{L}$ is non-degenerate, the second fundamental form of $\mathcal{L}$ is horizontal, i.e.*

$$h(X,Y) \in (Jz)^{\perp}, \forall X, Y \in T_z\mathcal{L}.$$

*Finally, the tensor $T(X,Y,Z) := \langle h(X,Y), JZ \rangle_{2p}$ is tri-symmetric.*

*Proof.* Let $X$ be a tangent vector field on a submanifold $\mathcal{L}$. The Legendrian assumption is equivalent to $\langle X, Jz \rangle_{2p} = 0, \forall z \in \mathcal{L}$. Differentiating this identity with respect to another tangent vector field $Y$ yields

$$\langle \nabla_Y X, Jz \rangle_{2p} + \langle X, JY \rangle_{2p} = 0. \tag{5.3}$$

Exchanging the roles of $X$ and $Y$ and subtracting, we get:

$$\langle [X,Y], Jz \rangle_{2p} + 2\langle X, JY \rangle_{2p} = 0.$$

The vector field $[X,Y]$ is tangent to $\mathcal{L}$, so by the Legendrian condition it belongs to $(Jz)^{\perp}$. Since $\langle [X,Y], Jz \rangle_{2p}$ vanishes, $\omega_p(Y,X) = \langle X, JY \rangle_{2p}$ vanishes as well. Next we have, by the definition of the second fundamental form and taking into account that $Jz$ is normal to $\mathcal{L}$:

$$\langle h(X,Y), Jz \rangle_{2p} = \langle \nabla_Y X, Jz \rangle_{2p}. \tag{5.4}$$

It then follows from Equation (5.3) that this expression vanishes, hence $h$ is horizontal. The proof of the fact that $T$ is tri-symmetric follows exactly the lines of the Lagrangian case and therefore is left to the reader. $\quad\square$

We now introduce, in analogy with the Lagrangian angle of Lagrangian submanifolds of Euclidean space, the Legendrian angle of a Legendrian submanifold:

**Definition 13.** *The Legendrian angle $\beta(z)$ of an oriented Legendrian submanifold of $\mathbb{Q}$ is the argument of $\Omega(z, X_1, ..., X_n)$, where $(X_1, ..., X_n)$ is an oriented basis of $T_z\mathcal{L}$. It is thus a map $\beta : \mathcal{L} \to \mathbb{R}/2\pi\mathbb{Z}$.*

**Proposition 12.** *Let $\mathcal{L}$ be a non-degenerate, Legendrian, oriented submanifold of $\mathbb{Q}$ with Legendrian angle $\beta$ and mean curvature vector $\vec{H}$. Then the following relation holds:*

$$n\vec{H} = J\nabla\beta.$$

*Proof.* The proof is very similar to the corresponding formula for Lagrangian submanifolds of Euclidean space (Proposition 24): set a point $e_0 := z$ of $\mathcal{L}$ and let $(e_1, ..., e_n)$ be an oriented, orthonormal frame of $\mathcal{L}$ defined in a neighbourhood of $z$. The Legendrian assumption and Lemma 20 imply that $(e_0, e_1, ..., e_n)$ is a Hermitian frame of $\mathbb{C}^{n+1}$, i.e.

$$[\langle\langle e_j, e_k\rangle\rangle_p]_{0 \le j,k \le n} = diag(1, \epsilon_1, ..., \epsilon_n).$$

Moreover, the second fundamental form is horizontal, so the mean curvature vector as well, and we have

$$\vec{H} \in (Jz)^{\perp} \cap N_z\mathcal{L} = Span(Je_1, ..., Je_n).$$

Hence the claimed formula is equivalent to

$$\langle n\vec{H}, Je_j\rangle_{2p} = \langle J\nabla\beta, Je_j\rangle_{2p}, \quad 1 \le j \le n,$$

i.e.

$$\langle n\vec{H}, Je_j\rangle_{2p} = d\beta(e_j).$$

Differentiating the identity $e^{i\beta} = \Omega(e_0, ..., e_n)$ with respect to the vector $e_j$, we obtain

$$id\beta(e_j)e^{i\beta} = \sum_{k=0}^{n} \Omega(e_0, ..., D_{e_j}e_k, ..., e_n)$$

$$= \Omega(e_j, e_1, ..., e_n) + \sum_{k=1}^{n} \Omega\left(e_0, ..., \sum_{l=0}^{n} \epsilon_l\langle\langle e_l, D_{e_j}e_k\rangle\rangle_p e_l, ..., e_n\right)$$

$$= \sum_{k=1}^{n} \Omega(e_0, ..., \epsilon_k\langle\langle e_k, D_{e_j}e_k\rangle\rangle_p e_k, ..., e_n)$$

$$= \sum_{k=1}^{n} \epsilon_k\langle\langle e_k, D_{e_j}e_k\rangle\rangle_p e^{i\beta},$$

hence

$$id\beta(e_j) = \sum_{k=1}^{n} \epsilon_k \langle\langle e_k, D_{e_j} e_k \rangle\rangle_p.$$

Recalling that $\langle\langle .,.\rangle\rangle_p = \langle .,.\rangle_{2p} - i\omega_p = \langle .,.\rangle_{2p} - i\langle J.,.\rangle_{2p}$, we have

$$\langle\langle e_k, D_{e_j} e_k\rangle\rangle_p = \langle e_k, D_{e_j} e_k\rangle_{2p} - i\langle Je_k, D_{e_j} e_k\rangle_{2p}.$$

Differentiating the relation $\langle e_k, e_k\rangle_{2p} = \epsilon_k$ in the direction $e_j$ yields $\langle e_k, D_{e_j} e_k\rangle_{2p} = 0$, so that, taking into account that $Je_j$ is a normal to $\mathcal{L}$, we get

$$\langle\langle e_k, D_{e_j} e_k\rangle\rangle_p = i\langle Je_k, D_{e_j} e_k\rangle_{2p} = i\langle Je_k, h(e_j, e_k)\rangle_{2p}.$$

We conclude, using the tri-symmetry of $T = \langle h(.,.), J.\rangle_{2p}$:

$$d\beta(e_j) = \sum_{k=1}^{n} \epsilon_k \langle Je_j, h(e_k, e_k)\rangle_{2p} = \langle Je_j, n\vec{H}\rangle_{2p}. \qquad \square$$

The next theorem shows the importance of Legendrian submanifolds of $\mathbb{Q}$: there are a concrete way of studying Lagrangian submanifolds of $\mathbb{CP}_p^n$:

**Theorem 28.** *Let* $\tilde{f} : \mathcal{M} \to \mathbb{Q}$ *be a Legendrian immersion. Then* $\pi \circ \tilde{f} :$ $\mathcal{M} \to \mathbb{CP}_p^n$ *is a Lagrangian immersion. Conversely, let* $f : \mathcal{M} \to \mathbb{CP}_p^n$ *be a Lagrangian immersion,* $\tilde{\mathcal{M}}$ *the universal covering of* $\mathcal{M}$, *and* $\tilde{\pi} : \tilde{\mathcal{M}} \to \mathcal{M}$ *the covering projection. Then there exists a Legendrian immersion* $\tilde{f} :$ $\tilde{\mathcal{M}} \to \mathbb{Q}$, *such that* $\pi \circ \tilde{f} = f \circ \tilde{\pi}$ *(we say that* $\tilde{f}$ *is a* Legendrian lift *of* $f$*). The lift* $\tilde{f}$ *is unique up to a rotation by* $e^{i\theta_0} Id$ *of* $\mathbb{C}^{n+1}$. *Finally, the immersions* $\tilde{f}$ *and* $f \circ \tilde{\pi}$ *of* $\tilde{\mathcal{M}}$ *have the same first fundamental and their second fundamental forms* $h_{\tilde{f}}$ *and* $h_f$ *are related by the equation*

$$d\pi(h_{\tilde{f}}(\tilde{X}, \tilde{Y})) = h_f(X, Y), \qquad (5.5)$$

*where* $X$ *and* $Y$ *are two vector fields tangent to* $\mathcal{L} := f(\mathcal{M})$. *Moreover,*

$$d\pi(\vec{H}_{\tilde{f}}(x)) = \vec{H}_f(\tilde{\pi}(x)), \quad \forall x \in \tilde{\mathcal{M}}, \qquad (5.6)$$

*where* $\vec{H}_{\tilde{f}}$ *and* $\vec{H}_f$ *denote the mean curvature vectors of* $\tilde{f}$ *and* $f$ *respectively. In particular, the immersion* $f$ *is minimal if and only if the immersion* $\tilde{f}$ *is minimal as well, i.e. it has constant Legendrian angle.*

**Example 11.** Since an alternated two-form vanishes on a one-dimensional space, a curve of a symplectic manifold of dimension 2 is always Lagrangian. Let $\gamma$ be a curve $\frac{1}{2}\mathbb{S}^2 \simeq \mathbb{CP}^1$ parametrized by $\gamma(t) = \frac{1}{2}(\sin\psi(t), \cos\psi(t)e^{i\varphi(t)})$, as in Section 3.3.4 of Chapter 3.

A Legendrian lift $\tilde{\gamma} = (\gamma_1, \gamma_2)$ of $\gamma$ must satisfy (see Section 4.3.1 of the previous chapter):

$$-|\gamma_1|^2 + |\gamma_2|^2 = \cos\psi \quad \text{and} \quad 2\gamma_1\bar{\gamma}_2 = \sin\psi e^{i\varphi}.$$

Hence it takes the form $\tilde{\gamma} = (\sin(\psi/2)e^{i\varphi_1}, \cos(\psi/2)e^{i\varphi_2})$, where $\varphi_1 - \varphi_2 = \varphi$. Moreover the Legendrian assumption yields

$$0 = \langle \tilde{\gamma}', J\tilde{\gamma}\rangle_0 = \varphi_1' \sin^2(\psi/2) + \varphi_2' \cos^2(\psi/2) = 0.$$

Combining this with $\varphi_1' - \varphi_2' = \varphi'$, we find

$$\varphi_1' = \cos^2(\psi/2)\varphi' \quad \text{and} \quad \varphi_2' = -\sin^2(\psi/2)\varphi'.$$

If we assume furthermore that $\gamma$ has constant curvature, it is, up to rotation, a horizontal circle: $\gamma(t) = \frac{1}{2}(\cos\psi_0, \sin\psi_0 e^{it})$, where $\psi_0$ is a real constant and a Legendrian lift of $\gamma$ is

$$\tilde{\gamma}(t) = (\sin(\psi_0/2)e^{i\cos^2(\psi_0/2)t}, \cos(\psi_0/2)e^{-i\sin^2(\psi_0/2)t}).$$

**Example 12.** Analogously, given a curve $\gamma$ of $\frac{1}{2}\mathbb{H}^2 \simeq \mathbb{CH}^1$ parametrized by $\gamma(s) = \frac{1}{2}(\cosh\psi(t), \sinh\psi(t)e^{i\varphi(t)})$ (see Section 3.3.5 of Chapter 3), a Legendrian lift $\tilde{\gamma}$ is a curve of $d\mathbb{S}^3$ parametrized by

$$\tilde{\gamma}(t) = (\sinh(\psi/2)e^{i\int_t(\cosh^2(\psi/2)\varphi')dt}, \cosh(\psi/2)e^{i\int_t(\sinh^2(\psi/2)\varphi')dt}).$$

**Remark 27.** Even if $f$ is an embedding, $\tilde{f}$ is not necessarily so, and may not be proper: in Example 11 above, the Legendrian curve $\tilde{\gamma}$ is closed if and only if $\tan^2(\psi_0/2)$ is rationally related to $2\pi$. Otherwise, $\tilde{\gamma}(\mathbb{R})$ is a dense subset of the torus $\sin(\psi_0/2)\,\mathbb{S}^1 \times \cos(\psi_0/2)\,\mathbb{S}^1$ of $\mathbb{S}^3$.

**Remark 28.** One can define locally a Lagrangian function $\beta_f$ on $\mathcal{L} = f(\mathcal{M})$ as follows: given a point $x$ of $\mathcal{M}$, there exists a neighbourhood of $x$ of the form $\tilde{\pi}(U)$, where $U$ is an open subset of $\tilde{\mathcal{M}}$ and $\tilde{\pi}|_U$ is one to one. Then we set $\beta_f(x) := \beta(\tilde{\pi}^{-1}(x)), \forall x \in \tilde{\pi}(U)$, where $\beta$ is the Legendrian angle of $\tilde{\mathcal{L}} = \tilde{f}(\tilde{\mathcal{M}})$. Observe that this angle is defined up to an additive constant, since the lift $\tilde{\mathcal{L}}$ of $\mathcal{L}$ is defined up to a rotation by $e^{i\theta_0}Id$. However, the gradient $\nabla\beta_{\tilde{f}}$ is well defined globally. Moreover from Equation (5.6) and Theorem 12 the formula $n\vec{H}_f = \mathbb{J}\nabla\beta_f$ holds. There is actually a more general, but more abstract definition of the Lagrangian angle of a Lagrangian submanifold in the general setting of Kähler-Einstein manifolds (see [Hélein, Romon (2002)]). Exercise 5 of Chapter 4 shows that the complex space forms are Kähler-Einstein.

*Proof of Theorem 28.* First observe that the Legendrian condition implies that at a point $z$ of $\tilde{\mathcal{L}}$ the linear space $d\pi(T_z\tilde{\mathcal{L}})$ of $T_{\pi(z)}\mathbb{CP}^n_p$ is $n$-dimensional. It follows that $\mathcal{L} = \pi(\tilde{\mathcal{L}})$ is a $n$-dimensional submanifold. Moreover, given two vectors fields $X$ and $Y$ tangent to $\mathcal{L}$, there exist locally two lifts $\tilde{X}$ of $\tilde{Y}$ tangent to $\tilde{\mathcal{L}}$, i.e. such that $d\pi(\tilde{X}) = X$ and $d\pi(\tilde{Y}) = Y$. It follows from the construction of the pseudo-Kähler structure on $\mathbb{CP}^n_p$ and from Lemma 20 that

$$\varpi(X,Y) = \omega_p(\tilde{X},\tilde{Y}) = 0,$$

i.e. $\mathcal{L}$ is Lagrangian.

Conversely, let $x$ be a point of $\mathcal{M}$. Its image $f(x)$ takes the form $\pi(z)$, where $z$ is a point of $\mathbb{Q}$. Hence locally one can construct a lift of $f$ to $\mathbb{Q}$. In order to extend such a lift globally, we consider the universal covering $\tilde{\mathcal{M}}$ of $\mathcal{M}$ and the canonical projection $\tilde{\pi} : \tilde{\mathcal{M}} \to \mathcal{M}$. Hence we can construct a global lift of $f$, i.e. a map $\bar{f} : \tilde{\mathcal{M}} \to \mathbb{Q}$ such that $\pi \circ \bar{f} = f \circ \tilde{\pi}$. Of course the immersion $\bar{f}$ is not necessarily Legendrian. Next introduce the one-form on $\tilde{\mathcal{M}}$ defined by

$$\alpha(X) := -\langle d\bar{f}_x(X), J\bar{f}(x)\rangle_{2p} = -\omega_p(d\bar{f}_x(X), \bar{f}(x)), \quad \forall X \in T_x\tilde{\mathcal{M}}.$$

It is easy to see that the fact that $\mathcal{L}$ is Lagrangian implies that $\alpha$ is closed: given two tangent vector fields $X$ and $Y$, we have

$$\begin{aligned}
d\alpha(X,Y) &= X(\alpha(Y)) - Y(\alpha(X)) + \alpha([X,Y])\\
&= -X(\omega_p(d\bar{f}(Y),\bar{f})) + Y(\omega_p(d\bar{f}(X),\bar{f})) + \omega_p([d\bar{f}(X),d\bar{f}(Y)],f)\\
&= -\omega_p(d^2\bar{f}(X,Y),\bar{f}) - \omega_p(d\bar{f}(Y),d\bar{f}(X)\\
&\quad + \omega_p(d^2\bar{f}(Y,X),\bar{f}) + \omega_p(d\bar{f}(X),d\bar{f}(Y))\\
&= 2\omega_p(d\bar{f}(X),d\bar{f}(Y))\\
&= 2\varpi\big(d\bar{f}(d\pi(X)),d\bar{f}(d\pi(Y))\big) = 0
\end{aligned}$$

(we have used the fact that the second derivative $d^2\bar{f}$ is bilinear and that $[d\bar{f}(X),d\bar{f}(Y)]$ is tangent to $\tilde{\mathcal{L}}$, hence orthogonal to $J\bar{f}$). Hence, $\tilde{\mathcal{M}}$ being simply connected, $\alpha$ admits a primitive $\theta$. Setting $\tilde{f} := e^{i\theta}\bar{f}$, we have

$$d\tilde{f} = id\theta e^{i\theta}\bar{f} + e^{i\theta}d\bar{f}.$$

Therefore

$$\begin{aligned}
\langle d\tilde{f}(X), J\tilde{f}_\theta\rangle_{2p} &= d\theta(X)\langle e^{i\theta}\bar{f}, e^{i\theta}\bar{f}\rangle_{2p} + \langle e^{i\theta}d\bar{f}(X), e^{i\theta}J\bar{f}\rangle_{2p}\\
&= d\theta + \langle d\bar{f}(X), J\bar{f}\rangle_{2p}\\
&= \alpha(X) + \omega_p(d\bar{f}(X),\bar{f})\\
&= 0,
\end{aligned}$$

so $\tilde{\mathcal{L}} := \tilde{f}(\tilde{\mathcal{M}})$ is Legendrian. Any other primitive of $\alpha$ takes the form $\theta + \theta_0$, and the resulting submanifold is $e^{i\theta_0}\tilde{\mathcal{L}}$. Moreover $\pi(\tilde{\mathcal{L}}) = \pi \circ \tilde{f}(\tilde{\mathcal{M}}) = f(\mathcal{M}) = \mathcal{L}$ so $\tilde{f}$ is a lift of $f$ as required.

Finally, given two vectors fields $X$ and $Y$ tangent to $\mathcal{M}$, there exist locally two lifts $\tilde{X}$ of $\tilde{Y}$ tangent to $\tilde{\mathcal{M}}$, i.e. such that $d\tilde{\pi}(\tilde{X}) = X$ and $d\tilde{\pi}(\tilde{Y}) = Y$ and we have

$$\langle d\tilde{f}(\tilde{X}), d\tilde{f}(\tilde{Y}) \rangle_{2p} = \mathbb{G}(df(X), df(Y)),$$

proving that the first fundamental forms of $f$ and $\tilde{f}$ are equal.

Moreover, using Proposition 4.7 of Chapter 4, we obtain

$$d\pi(D_{\tilde{Y}}\tilde{X}) = \nabla_Y X.$$

On the other hand, given a point $z$ of $\tilde{\mathcal{L}}$, we have $d\pi(T_z\tilde{\mathcal{L}}) = T_{\pi(z)}\mathcal{L}$. These two facts imply that

$$\begin{aligned} d\pi(h_{\tilde{f}}(\tilde{X}, \tilde{Y})) &= d\pi((D_{\tilde{Y}}\tilde{X})^\perp) \\ &= (\nabla_Y X)^\perp \\ &= h_f(X, Y). \end{aligned}$$

From the definition of the mean curvature vector, and the fact that the first fundamental forms of $f$ and $\tilde{f}$ are identical, it follows that $d\pi(\vec{H}_{\tilde{f}}) = \vec{H}_f$. $\qquad\square$

### 5.5.2 *Equivariant Legendrian submanifolds in odd-dimensional space forms*

Consider again the action of $SO(p, n+1-p)$ on $\mathbb{C}^{n+1}$ introduced in Section 5.4.2. Given a matrix $M \in SO(p, n+1-p)$, we have $z \sim z' \Leftrightarrow Mz \sim Mz'$. Therefore the action of $SO(p, n+1-p)$ may be "projected" on $\mathbb{CP}^n$. Moreover, since this action preserves the causal character, it induces an action on $\mathbb{CP}^n_p$. It is easy to check that the orbit of a point $z$ of $\mathbb{Q}^{2n+1}_{2p,1}$ by the $SO(p, n+1-p)$ action is the Legendrian quadric

$$\mathbb{Q}^n_{p,1}z := \left\{ zx \,\middle|\, x \in \mathbb{Q}^n_{p,1} = \mathbb{Q}^{2n+1}_{2p,1} \cap \mathbb{R}^{n+1} \right\}.$$

We leave to the reader the easy task of checking that this quadric is minimal (and even totally geodesic). Hence the orbits of the corresponding action in $\mathbb{CP}^n_p$ are the totally geodesic, Lagrangian submanifolds $\pi(\mathbb{Q}^n_{p,1}z)$. Following the lines of Section 3.3 of Chapter 3, we introduce a subgroup of $SO(p, n +$

$1 - p)$ (and therefore of $SU(p, n + 1 - p)$) whose orbits (both in $\mathbb{Q}^{2n+1}_{2p,1}$ and in $\mathbb{CP}^n_p$) are $(n - 1)$-dimensional submanifolds:

$$\widetilde{SO}(p, n - p) := \left\{ \begin{pmatrix} M & 0 \\ 0 & 1 \end{pmatrix} \,\middle|\, M \in SO(p, n - p) \right\}.$$

The next theorem characterizes those Lagrangian and Legendrian submanifolds which are invariant with respect to the action of $\widetilde{SO}(p, n - p)$.

**Theorem 29.** *Let $\mathcal{L}$ be a Lagrangian submanifold in $\mathbb{CP}^n_p$ which is $SO(p, n - p)$-equivariant. Then its Legendrian lift $\tilde{\mathcal{L}}$ is $SO(p, n - p)$-equivariant as well. Moreover, $\tilde{\mathcal{L}}$ may be locally parametrized by an immersion of the form*

$$\begin{aligned} f : I \times \mathbb{Q}^{n-1}_{p,\epsilon} &\to \quad \mathbb{Q}^{2n+1}_{2p,1} \\ (s, x) &\mapsto (\gamma_1(s)x, \gamma_2(s)), \end{aligned}$$

*where $\tilde{\gamma} := (\gamma_1, \gamma_2) : I \to \mathbb{Q}^3_{1-\epsilon,1}$ is a Legendrian curve. In other words, $\tilde{\gamma}$ is a Legendrian curve of $\mathbb{Q}^3_{0,1} = \mathbb{S}^3$ if $\epsilon = 1$ and of $\mathbb{Q}^3_{2,1} = Ad\mathbb{S}^3$ if $\epsilon = -1$. Moreover, the Legendrian angle of $\tilde{\mathcal{L}}$ is given by*

$$\beta_{\tilde{\mathcal{L}}} = \beta_{\tilde{\gamma}} + (n - 1) \arg \gamma_1,$$

*where $\beta_{\tilde{\gamma}}$ denotes the Legendrian angle of $\tilde{\gamma}$.*

*Proof.* The proof is similar to the one of Theorem 26 (Section 5.4.2):

**First case:** $n = 2$. Recall that the Hermitian form on $\mathbb{C}^3$ is

$$\langle\langle ., . \rangle\rangle_p = \epsilon_1 dz_1 d\bar{z}_1 + \epsilon_2 dz_2 d\bar{z}_2 + \epsilon_3 dz_3 d\bar{z}_3.$$

We set $\eta := \epsilon_1 \epsilon_2$ and

$$M_\eta := \begin{pmatrix} 0 & -\eta & 0 \\ 1 & 0 & 0 \\ 0 & 0 & 0 \end{pmatrix},$$

so that

$$\widetilde{SO}(2) = \{ e^{M_1 t}, t \in \mathbb{R} \} \quad \text{and} \quad \widetilde{SO}(1, 1) = \{ e^{M_{-1} t}, t \in \mathbb{R} \}.$$

A surface of $\mathbb{Q}^5_{2p,1}$ which is $SO(2)$ or $SO(1, 1)$-equivariant may be locally parametrized by an immersion of the form:

$$f(s, t) = e^{M_\eta t}(z_1(s), z_2(s), z_3(s)),$$

where $(z_1(s), z_2(s), z_3(s))$ is some curve in $\mathbb{Q}^5_{2p,1}$ We first compute the first derivatives of the immersion:

$$\begin{aligned} f_s &= e^{M_\eta t}(z'_1, z'_2, z'_3) \\ f_t &= e^{M_\eta t} M_\eta(z_1, z_2, z_3) = e^{M_\eta t}(-\eta z_2, z_1, 0). \end{aligned}$$

Therefore the Legendrian condition yields:

$$0 = \langle f_t, Jf \rangle_{2p} = -\epsilon_1 \eta \langle z_2, i z_1 \rangle + \epsilon_2 \langle z_1, i z_2 \rangle = 2\epsilon_2 \langle z_1, i z_2 \rangle.$$

Thus, $z_1$ and $z_2$ have the same argument. In polar coordinates, $z_1 = r_1 e^{i\varphi}$ and $z_2 = r_2 e^{i\varphi}$. In order to complete the case $n = 2$, we need to consider separately the definite and indefinite cases:

$\eta = 1$: *The $SO(2)$ case:* Observe that the second coordinate of $f$ is $z_2(s)\cos t + z_1(s)\sin t = (r_2(s)\cos t + r_1(s)\sin t)e^{i\varphi(s)}$. Clearly, for all $s \in I$, there exists $t(s)$ such that $r_2(s)\cos t(s) + r_1(s)\sin t(s) = 0$, hence the second coordinate of $f$ vanishes at $(s, t(s))$. Setting $\gamma_1(s) := z_1(s)\cos t(s) - z_2(s)\sin t(s)$, i.e. $\gamma_1(s)$ is the first coordinate of $f$ at $(s, t(s))$, we see that $f(s,t) = e^{M_n(t-s(t))}(\gamma_1(s), 0, \gamma_2(s))$, where we set $\gamma_2(s) := z_3(s)$. Hence the immersion $\tilde{f}(s,t) := e^{M_n t}(\gamma_1(s), 0, \gamma_2(s)) = (\gamma_1 \cos t, \gamma_1 \sin t, \gamma_2)$ parametrizes the same surface and we get the required parametrization of our surface.

$\eta = -1$: *The $SO(1,1)$ case:* We first observe that $r_1 \neq r_2$ since otherwise the immersion would be degenerate. If $r_1 > r_2$, there exists $t(s)$ such that $r_2(s)\cosh t(s) + r_1(s)\sinh t(s) = 0$, hence the second coordinate of $f$ vanishes at $(s, t(s))$. Similarly to the positive case, we set $\gamma(s) = r_1(s)\cosh t(s) + r_2(s)\sinh t(s)$, and as before, we check that $\tilde{f}(s,t) := (\gamma_1(s)\cosh t, \gamma_1(s)\sinh t, \gamma_2(s))$, where we set $\gamma_2(s) := z_3(s)$, parametrizes the same surface than $f$. The argument is strictly analogous if $r_1 < r_2$: we find $t(s)$ in order to make the first coordinate vanish and find $\tilde{f}(s,t) = (\gamma_1(s)\sinh t, \gamma_1(s)\cosh t, \gamma_2(s))$.

In both cases, the fact that $f$ is Legendrian (precisely the vanishing of $\langle f_s, Jf \rangle_{2p}$) implies that $\tilde{\gamma}$ is Legendrian as well.

**Second case: $n \geq 3$.**

Let $z$ be a point of $\tilde{\mathcal{L}}$. We claim that there exists $x \in \mathbb{Q}_{p,\epsilon}^{n-1}$ and $\tilde{\gamma} = (\gamma_1, \gamma_2) \in \mathbb{Q}_{1-\epsilon}^3$ such that $z = (\gamma_1 x, \gamma_2)$. To prove this, recall that the Hermitian form on $\mathbb{C}^{n+1}$ is

$$\langle\langle .,. \rangle\rangle_p = \sum_{j=1}^{n+1} \epsilon_j dz_j d\bar{z}_j$$

and consider the two skew matrices $M_{jl}$ and $M_{kl}$, with $1 \leq j, k, l \leq n$, and $j \neq k \neq l \neq j$, defined by

$$M_{jl}e_j = e_l, \quad M_{jl}e_l = \epsilon_j \epsilon_l e_j \quad \text{and} \quad M_{jl}e_m = 0 \quad \text{for } m \neq j, l$$

and

$$M_{kl}e_k = e_l, \quad M_{kl}e_l = \epsilon_k\epsilon_l e_k \text{ and } M_{kl}e_m = 0 \text{ for } m \neq k, l.$$

The reader may check that the curves $s \mapsto e^{M_{jl}s}$ and $s \mapsto e^{M_{kl}s}$ belong to $\widetilde{SO}(p, n-p)$. Hence, given a point $z$ in $\mathcal{L}$, the two curves $s \mapsto e^{M_{jl}s}z$ and $s \mapsto e^{M_{kl}s}z$ belong to $\mathcal{L}$ and the two vectors $M_{jl}z$ and $M_{kl}z$ are tangent to $\mathcal{L}$ at $z$. Thus the Legendrian assumption yields

$$0 = \omega_p(M_{jl}z, M_{kl}z) = \text{Re } z_j \text{Im } z_k - \text{Re } z_k \text{Im } z_j.$$

Since this holds for any pair of indexes $(j, k)$, it follows that $(\text{Re } z_1, ..., \text{Re } z_n)$ and $(\text{Im } z_1, ..., \text{Im } z_n)$ are collinear. Therefore there exists $\gamma_1 \in \mathbb{C}$ and $x \in \mathbb{Q}_{p,\epsilon}^{n-1}$ such that $(z_1, ..., z_n) = \gamma_1 x$. The equivariant assumption implies that the quadric $(\gamma_1 \mathbb{Q}_{p,\epsilon}^{n-1}, \gamma_2)$, where we set $\gamma_2 := z_{n+1}$, is contained in $\tilde{\mathcal{L}}$. Hence the latter, being $n$-dimensional, is locally foliated by a one-parameter family of quadrics $(\gamma_1(s)\mathbb{Q}_{p,\epsilon}^{n-1}, \gamma_2(s))$, hence it is parametrized by an immersion of the form $f(s, x) = (\gamma_1(s)x, \gamma_2(s))$. The assumption $f(s, x) \in \mathbb{Q}$ implies that the curve $\tilde{\gamma} = (\gamma_1, \gamma_2)$ is contained in $\mathbb{S}^3$ if $\epsilon = 1$ and in $Ad\mathbb{S}^3$ if $\epsilon = -1$, and the fact that $f$ is Legendrian implies that $\gamma$ is Legendrian as well.

We now prove the second part of the theorem: let $f$ be an immersion as described in the statement of the theorem, $x$ a point of $\mathbb{Q}_{p,\epsilon}^{n-1}$ and $(e_1, ..., e_{n-1})$ an oriented orthonormal basis of $T_x\mathbb{Q}_{p,\epsilon}^{n-1}$. Setting

$$X_j := (\gamma_1 e_j, 0) \quad \text{and} \quad X_n := (\gamma_1'x, \gamma_2'),$$

it is easy to check that $(X_1, ..., X_n)$ is a basis of $T_z\tilde{\mathcal{L}}$ at the point $z := (\gamma_1 x, \gamma_2)$. Then we calculate

$$\langle X_j, Jz \rangle_{2p} = \langle \gamma_1, i\gamma_1 \rangle \langle e_j, z \rangle_p = 0,$$
$$\langle X_n, Jz \rangle_{2p} = \epsilon\langle \gamma_1, i\gamma_1' \rangle + \langle \gamma_2, i\gamma_2' \rangle = 0,$$

(we have used the fact that the curve $\gamma$ is Legendrian in the second line), which shows that $\tilde{\mathcal{L}}$ is Legendrian. Finally, its Legendrian angle is obtained from the following computation:

$$\begin{aligned}
e^{i\beta_{\tilde{\mathcal{L}}}} &= \Omega(z, X_1, ..., X_n) \\
&= \Omega\Big((\gamma_1 x, \gamma_2)(\gamma_1 e_1, 0), ..., (\gamma_1 e_{n-1}, 0), (\gamma_1'x, \gamma_2')\Big) \\
&= \gamma_1^{n-1} \det(e_1, ..., e_{n-1}, x)(\gamma_1\gamma_2' - \gamma_2\gamma_1') \\
&= \gamma_1^{n-1} \det_{\mathbb{C}}(\tilde{\gamma}, \tilde{\gamma}') = \gamma_1^{n-1}e^{i\beta_{\tilde{\gamma}}}.
\end{aligned}$$

$\square$

It turns out that it is not difficult to extend our study to those Legendrian submanifolds which are equivariant by the action of the subgroup of $SO(p, n+1-p)$ defined as follows:

$$\left\{ \begin{pmatrix} M & 0 \\ 0 & N \end{pmatrix} \middle| M \in SO(p_1, n_1 - p_1), N \in SO(p_2, n_2 - p_2) \right\},$$

where $n_1 + n_2 = n + 1$ and $p_1 + p_2 = p$.

Hence Theorem 29 is a particular case of the following one:

**Theorem 30.** *Let $\mathcal{L}$ be a Lagrangian submanifold in $\mathbb{CP}_p^n$ which is $SO(p_1, n_1 - p_1) \times SO(p_2, n_2 - p_2)$-equivariant with $p_1 + p_2 = p$ and $n_1 + n_2 = n + 1$. Then its Legendrian lift $\tilde{\mathcal{L}}$ is $SO(p_1, n_1 - p_1) \times SO(p_2, n_2 - p_2)$-equivariant as well. Moreover, $\tilde{\mathcal{L}}$ may be locally parametrized by an immersion of the form*

$$\begin{aligned} f : I \times \mathbb{Q}_{2p_1, \epsilon}^{n_1 - 1} \times \mathbb{Q}_{2p_2, 1}^{n_2 - 1} &\to \qquad \mathbb{Q}_{2p, 1}^{2n+1} \\ (s, x, y) &\mapsto (\gamma_1(s)x, \gamma_2(s)y), \end{aligned}$$

*where $\tilde{\gamma} := (\gamma_1, \gamma_2)$ is a Legendrian curve of $\mathbb{Q}_{1-\epsilon, 1}^3$. In other words, $\tilde{\gamma}$ is a Legendrian curve of $\mathbb{S}^3$ if $\epsilon = 1$ and of $d\mathbb{S}^3$ if $\epsilon = -1$. Moreover, the Legendrian angle of $\tilde{\mathcal{L}}$ is given by*

$$\beta_{\tilde{\mathcal{L}}} = \beta_{\tilde{\gamma}} + (n_1 - 1) \arg \gamma_1 + (n_2 - 1) \arg \gamma_2, \tag{5.7}$$

*where $\beta_{\tilde{\gamma}}$ denotes the Legendrian angle of $\tilde{\gamma}$.*

*Proof.* The proof requires to consider a number of cases, depending whether $n_1$ and $n_2$ are equal or greater than 3, but follows closely that of Theorem 29. The details are therefore left to the Reader. $\square$

### 5.5.3 *Minimal equivariant Lagrangian submanifolds in complex space forms*

#### 5.5.3.1 *The minimal equation*

**Lemma 21.** *The immersion $f$ of Theorem 30 is minimal if and only if the curve $\pi(\tilde{\gamma}) := \frac{1}{2}(r, x_1, x_2)$ of $\mathbb{Q}_{1-\epsilon, 1/4}^2$ is solution of the EDO:*

$$\kappa + \nu_1 \left( \frac{n_1 - 1}{1 - r} - \frac{n_2 - 1}{1 + r} \right) = 0, \tag{5.8}$$

*where $\nu = (\nu_1, \nu_2, \nu_3)$ is a unit normal vector to $\gamma := \pi(\tilde{\gamma})$ and $\kappa$ denotes the curvature of $\gamma$ with respect to $\nu$.*

*Proof.* By Theorem 30, the immersion $f$ is minimal if and only if $\beta_{\tilde{L}}$ is constant. The proof therefore consists of differentiating Equation (5.7) of Theorem 30. In the following we set $\beta := \beta_{\tilde{\gamma}}$ and let $s$ be the arc length parameter on $\tilde{\gamma}$. Then it is also the arc length parameter of $\gamma := \pi(\tilde{\gamma})$. The proof is divided into several steps:

*First step:* $\kappa = \frac{d\beta}{ds}$.
This is actually the one-dimensional version of the general formula (5.6) of Theorem 28. To be more rigorous, observing that $|\gamma|^2_{1-\epsilon} = |\tilde{\gamma}'|^2_{1-\epsilon} = \epsilon$, we have on the one hand

$$\nabla\beta = \epsilon\beta'\tilde{\gamma}'$$

and on the other one

$$\vec{H}_{\tilde{\gamma}} = \epsilon\tilde{\gamma}'',$$

so that, from the formula $H = J\nabla\beta$ , we obtain $\tilde{\gamma}'' = \beta'J\tilde{\gamma}'$. Projecting this equation by $d\pi$, we get $\gamma'' = d\pi(\tilde{\gamma}'') = \beta'j\gamma'$. Therefore, by the Frénet equation, and remembering that the metric on $\mathbb{Q}^2_{1-\epsilon,1/4}$ is definite, the curvature of $\gamma$ with respect to the unit normal vector $\nu = j\gamma'$ is $\kappa = \beta'$.

*Second step:* $(\arg\gamma_1)' = \frac{\langle\gamma_1', i\gamma_1\rangle}{|\gamma_1|^2}$.
Differentiate the identity $\gamma_1 = |\gamma_1|e^{i\arg\gamma_1}$ to obtain

$$\gamma_1' = (|\gamma_1|' + i|\gamma_1|(\arg\gamma_1)')e^{i\arg\gamma_1}.$$

Therefore $\langle\gamma_1', i\gamma_1\rangle = |\gamma_1|^2(\arg\gamma_1)'$, hence the claim. Of course, we have $(\arg\gamma_2)' = \frac{\langle\gamma_2', i\gamma_2\rangle}{|\gamma_2|^2}$ as well.

*Third step:* $\nu_1 = 2\epsilon\langle\gamma_1', i\gamma_1\rangle = -2\langle\gamma_2', i\gamma_2\rangle$, $|\gamma_1|^2 = \frac{\epsilon(1-r)}{2}$ and $|\gamma_2|^2 = \frac{1+r}{2}$.
Recall that $\nu = j\gamma' = \epsilon\,2\gamma\times\gamma'$, and $\langle u\times v, w\rangle_{1-\epsilon} = \det(u, v, w)$. We also use the Legendrian assumption $\epsilon\langle\gamma_1', i\gamma_1\rangle + \langle\gamma_2', i\gamma_2\rangle = 0$. Hence we calculate

$$\begin{aligned}
\epsilon\nu_1 &= \langle j\gamma', (1,0,0)\rangle_{1-\epsilon} = 2\det(\gamma, \gamma', (1,0,0))\\
&= 2\det(\gamma_1\bar{\gamma}_2, (\gamma_1\bar{\gamma}_2)')\\
&= 2\langle i\gamma_1\bar{\gamma}_2, \gamma_1'\bar{\gamma}_2 + \gamma_1\bar{\gamma}_2'\rangle\\
&= 2|\gamma_2|^2\langle\gamma_1', i\gamma_1\rangle + 2|\gamma_1|^2\langle\bar{\gamma}_2', i\bar{\gamma}_2\rangle\\
&= 2|\gamma_2|^2\langle\gamma_1', i\gamma_1\rangle - 2|\gamma_1|^2\langle\gamma_2', i\gamma_2\rangle\\
&= 2|\gamma_2|^2\langle\gamma_1', i\gamma_1\rangle + 2\epsilon|\gamma_1|^2\langle\gamma_1', i\gamma_1\rangle\\
&= 2\langle\gamma_1', i\gamma_1\rangle = -2\epsilon\langle\gamma_2', i\gamma_2\rangle.
\end{aligned}$$

Finally we get the expression of $|\gamma_1|^2$ and $|\gamma_2|^2$ solving the system

$$\begin{cases} \epsilon|\gamma_1|^2 + |\gamma_2|^2 = 1\\ -\epsilon|\gamma_1|^2 + |\gamma_2|^2 = r. \end{cases}$$

Gathering all these facts and differentiating Equation (5.7), we get Equation (5.8). □

We now show that, as in Chapter 3, Equation (3.6) enjoys a first integral; due to the form of the equation, we look for a function of the form $E(\gamma, \gamma') = E(r, \nu_1) = \nu_1 F(r)$, where $F$ is some real-valued function. Therefore we have

$$\frac{d}{ds}E = \nu_1'F(r) + \nu_1 r'\frac{dF(r)}{dr} = r'F(r)\left(-\kappa + \nu_1 F^{-1}(r)\frac{dF(r)}{dr}\right),$$

so, in view of Equation (3.6), $F$ must satisfy

$$\frac{d}{dr}\log(F(r)) = \frac{n_2 - 1}{1 + r} - \frac{n_1 - 1}{1 - r}.$$

Hence a suitable choice for $F$ is $F(r) = (1 - r)^{n_1 - 1}(1 + r)^{n_2 - 1}$.

### 5.5.3.2  *Special solutions*

As in Section 3.3.3, we first discuss the existence of special solutions of Equation (5.8). Again, the case $E = 0$ implies the vanishing of $\nu_1$ and $\kappa$, so the corresponding curve $\gamma$ is a geodesic and its Legendrian lift $\tilde{\gamma}$ as well. It is then a simple exercise to check that the corresponding Legendrian submanifold $\tilde{\mathcal{L}}$ takes the form $\tilde{\mathcal{L}} = \mathbb{Q} \cap E$, where $E$ is a Lagrangian subspace of $\mathbb{C}^{n+1}$. In particular $\tilde{\mathcal{L}}$ is totally geodesic.

We look for other special solutions assuming that $r = \bar{r}$ is constant. An easy computation shows that there is no such solution in the hyperbolic case, and that $\bar{r} := \frac{n_1 - n_2}{n}$ is the unique one in the spherical case. The corresponding curve $\gamma$ is the horizontal circle of height $\bar{r}$ as described in Example 11. Observe that its Legendrian lift $\tilde{\gamma}$ is closed as well, since $\frac{1 - \bar{r}^2}{\bar{r}^2}$ is rational (see Remark 27). We may actually give an explicit parametrization of $\tilde{\gamma}$:

$$\tilde{\gamma}(t) = \left(\sqrt{\frac{2n_1 - 1}{2n}}e^{i\frac{2n_2 - 1}{2n}t}, \sqrt{\frac{2n_2 - 1}{2n}}e^{-i\frac{2n_1 - 1}{2n}t}\right).$$

### 5.5.3.3  *The spherical case $\epsilon = 1$*

As in Section 3.3.4 of Chapter 3, we use spherical coordinates on $\mathbb{S}^2$:

$$\gamma = \frac{1}{2}(r, z_1, z_2) = \frac{1}{2}(\cos\psi, \sin\psi\cos\varphi, \sin\psi\sin\varphi),$$

where $\psi \in (0, \pi)$ and $\varphi \in \mathbb{R}/2\pi\mathbb{Z}$, and the orthonormal frame

$$e_1 := (0, -\sin\varphi, \cos\varphi) \quad \text{and} \quad e_2 := (-\sin\psi, \cos\psi\cos\varphi, \cos\psi\sin\varphi).$$

In particular, $\nu_1 = \sin\psi\cos\theta$, so

$$E = \cos\theta\sin\psi\, F(\cos\psi) = \cos\theta\, e(\psi),$$

where we set

$$e(\psi) := \sin\psi\, F(\cos\psi) = 2^{n_1+n_2-1}\sin^{2n_1-1}(\psi/2)\cos^{2n_2-1}(\psi/2).$$

The projections of the solutions $(\gamma,\gamma')$ on the plane $(\theta,\psi)$ are of three types:

(i) the vertical segment $\psi = 0[\pi]$ with energy level $E = 0$;

(ii) the point $(0[\pi],\psi_0)$, where $\psi_0 := \arccos(\frac{n_1-n_2}{n}) = 2\arccos(\sqrt{\frac{2n_1-1}{2n}})$. Its energy level is $E_0 := e(\psi_0)$ and the corresponding curve $\gamma$ is a horizontal circle;

(iii) a closed curve winding around the fixed point $(0[\pi],\psi_0)$ and contained in the open subset

$$\{-\pi/2 < \theta < \pi/2, 0 < \psi < \pi\}.$$

The solutions (i) and (ii) have been discussed in the section above. As in Section 3.3.4 of Chapter 3, a solution of type (iii) gives rise to a closed curve $\gamma$ is and only if the variation of the angle $\varphi$ over a period $T$ of the angle $\psi$, i.e. the quantity $\Phi(E) := \int_s^{s+T}\varphi'(\sigma)d\sigma$ takes the form $\frac{a}{b}2\pi$. Observe that even if $\gamma$ is closed, its Legendrian lift is not necessarily so, and therefore the corresponding Legendrian submanifold $\tilde{\mathcal{L}}$ may not be properly immersed. We claim however that if $\gamma$ is closed, then $\mathcal{L}$ is properly immersed. To see this, consider two real numbers $s_1$ and $s_2$ such that $\gamma(s_1) = \gamma(s_2)$. It follows that there exists a real $\theta$ such that $\tilde{\gamma}(s_1) = e^{i\theta}\tilde{\gamma}(s_2)$. Therefore, for all $(x,y) \in \mathbb{Q}_{p_1,\epsilon}^{n_1-1}\times\mathbb{Q}_{p_2,1}^{n_2-1}$, we have

$$\pi(\gamma_1(s_1)x,\gamma_2(s_1)y) = \pi(e^{i\theta}(\gamma_1(s_2)x,\gamma_2(s_2)y)) = \pi(\gamma_1(s_2)x,\gamma_2(s_2)y).$$

Hence the one-parameter family of orbits $\pi(\gamma_1(s)\mathbb{Q}_{p_1,\epsilon}^{n_1-1},\gamma_2(s)\mathbb{Q}_{p_2,1}^{n_2-1})$ closes up if and only if $\gamma$ closes up. The next lemma shows that there is a countable family of such curves.

**Lemma 22.** *The variation of $\varphi$ on a period of $\psi$, i.e. the number $\Phi(E)$ ranges a non-empty interval. Hence there is a countable family of closed solutions of Equation (3.6). The other solutions are non properly immersed curves.*

*Proof.* The proof follows from the fact that $\lim_{E\to 0}\Phi(E) \neq \lim_{E\to E_0}\Phi(E)$. Therefore the number $\Phi(E)$ ranges at least the inverval bounded by these two limits.

Denote by $\psi_-(E)$ and $\psi_+(E)$ the extremal values taken by $\psi$ on an integral curve of energy level $E$, or, equivalently, the two roots of the equation $e(\psi) = E$. Observe that $\lim_{E \to E_0} \psi_-(E) = \lim_{E \to E_0} \psi_+(E) = \psi_0$. By the symmetry of the equation, we have

$$\Phi(E) = 2 \int_{\psi_-}^{\psi_+} \frac{\varphi'}{\psi'} d\psi = 2(\Phi_-(\psi_-) + \Phi_+(\psi_+)),$$

where we set

$$\Phi_-(\psi_-) := \int_{\psi_-}^{\psi_0} \frac{\cot\theta}{\sin\psi} d\psi \quad \text{and} \quad \Phi_+(\psi_+) := \int_{\psi_0}^{\psi_+} \frac{\cot\theta}{\sin\psi} d\psi.$$

By the definition of $e(\psi)$, we have

$$\Phi_+(\psi_+) = \int_{\psi_0}^{\psi_+} \left( \frac{e^2(\psi)}{e^2(\psi_+)} - 1 \right)^{-1/2} \frac{d\psi}{\sin\psi}.$$

Setting $h := \psi_+ - \psi_0$ and making the change of variable $\psi = \psi_0 + hx$, we get

$$\Phi_+(\psi_+) = \int_0^1 \left( \frac{e^2(\psi_0 + hx)}{e^2(\psi_0 + h)} - 1 \right)^{-1/2} \frac{h\,dx}{\sin(\psi_0 + hx)}.$$

We introduce

$$f(\psi) := \frac{1}{2} \left( (2n_1 - 1) \cot(\psi/2) - (2n_2 - 1) \tan(\psi/2) \right),$$

so that $e'(\psi) = f(\psi)e(\psi)$. Therefore, using the facts that $f(\psi_0) = 0$ and $e''(\psi) = (f'(\psi) + f^2(\psi))e(\psi)$, we calculate

$$e''(\psi_0) = f'(\psi_0)e(\psi_0) = -\frac{n_1 + n_2 - 1}{2}e(\psi_0) = -\frac{n}{2}e(\psi_0).$$

Hence we get the following Taylor expansion:

$$\left( \frac{e(\psi_0 + hx)}{e(\psi_0 + h)} \right)^2 - 1 = \left( \frac{1 + \frac{1}{2}f'(\psi_0)(hx)^2 + o(h^2)}{1 + \frac{1}{2}f'(\psi_0)h^2 + o(h^2)} \right)^2 - 1$$

$$= \frac{n}{2}h^2(1 - x^2) + o(h^2).$$

On the other hand

$$\sin(\psi_0 + hx) = \sin(\psi_0) + o(h) = \frac{\sqrt{(2n_1 - 1)(2n_2 - 1)}}{n} + o(h),$$

so finally

$$\Phi_+(\psi_+) = \sqrt{\frac{2n}{(2n_1 - 1)(2n_2 - 1)}} \int_0^1 \frac{dx}{\sqrt{1 - x^2}} + o(h),$$

which implies

$$\lim_{\psi_+ \to \psi_0} \Phi_+(\psi_+) = \sqrt{\frac{2n}{(2n_1 - 1)(2n_2 - 1)}} \pi/2.$$

A similar computation shows that $\lim_{\psi_- \to \psi_0} \Phi_-(\psi_-) = \lim_{\psi_+ \to \psi_0} \Phi_+(\psi_+)$ so we get

$$\lim_{E \to E_0} \Phi(E) = \sqrt{\frac{2n}{(2n_1 - 1)(2n_2 - 1)}} 2\pi = \sqrt{\frac{1}{2n_1 - 1} + \frac{1}{2n_2 - 1}} 2\pi.$$

In order to compute the limit of $\Phi_+(\psi_+)$ at $\psi_+ \to \pi$ (i.e. at $E \to 0$), we make the following change of variable: $\psi = \pi - hx$, where we set $h := \pi - \psi_+$. The inverse formula is $x = \frac{\pi - \psi}{\pi - \psi_+} = \frac{\pi - \psi}{h}$. Hence we get

$$\Phi_+(\psi_+) = \int_1^{(\pi - \psi_0)/h} \left( \frac{e^2(\pi - hx)}{e^2(\pi - h)} - 1 \right)^{-1/2} \frac{h \, dx}{\sin(\pi - hx)}.$$

It follows that

$$\lim_{\psi_+ \to \pi} \Phi_+(\psi_+) = \int_1^\infty \frac{dx}{x\sqrt{x^{2n_2 - 1} - 1}} = \frac{\pi}{2n_2 - 1}.$$

Analogously we show that

$$\lim_{\psi_- \to 0} \Phi_+(\psi_-) = \int_1^\infty \frac{dx}{x\sqrt{x^{2n_1 - 1} - 1}} = \frac{\pi}{2n_1 - 1}.$$

We conclude that

$$\lim_{E \to 0} \Phi(E) = \left( \frac{1}{2n_1 - 1} + \frac{1}{2n_2 - 1} \right) 2\pi. \qquad \square$$

**Remark 29.** Observe that if $n_1 > 1$ and $n_2 > 1$, then $\lim_{E \to 0} \Phi(E) < \lim_{E \to E_0} \Phi(E) < 2\pi$, while the reverse inequalities hold if $n_1 = 1$ or $n_2 = 1$.

### 5.5.3.4 *The hyperbolic case $\epsilon = -1$*

As in Section 3.3.5, we introduce "elliptic" coordinates on the hyperbolic plane

$$\gamma = \frac{1}{2}(r, z_1, z_2) = \frac{1}{2}(\cosh \psi, \sinh \psi e^{i\varphi}),$$

where $\psi \in (0, \infty), \varphi \in \mathbb{R}/2\pi\mathbb{Z}$. The curves projections of the curves $(\gamma, \gamma')$ on the plane $(\theta, \psi)$ have the same profile than those of Section 3.3.5, and we obtain by a similar computation that the total variation of the angle $\varphi$ is bounded by $\frac{\pi}{n_1 - 1}$. Therefore the curves $\gamma$ are embedded, have two unbounded ends, and are contained in an angular sector $\{\varphi_0 \leq \varphi \leq \varphi_0 + \frac{\pi}{n_1 - 1}\}$ of the hyperbolic plane.

### 5.5.3.5 *Conclusion*

**Theorem 31.** *Let $\mathcal{L}$ be a connected, minimal Lagrangian of the complex space form $\mathbb{CP}_p^n$ which is $SO(p_1, n_1 - p_1) \times SO(p_2, n_2 - p_2)$-equivariant with $p_1 + p_2 = p$ and $n_1 + n_2 = n + 1$. Then its Legendrian lift $\tilde{\mathcal{L}}$ is congruent to an open subset of one the following*

(i) *a totally geodesic Legendrian quadric, i.e. $\tilde{\mathcal{L}} = \mathbb{Q} \cap E$, where $E$ is a Lagrangian subspace of $\mathbb{C}^{n+1}$. In particular $\mathcal{L}$ is totally geodesic;*

(ii) *the minimal Legendrian submanifold*

$$\left\{ (ae^{ib^2t}x, be^{-ia^2t}y) \in \mathbb{Q}_{2p,1}^{2n+1} \,\middle|\, (t, x, y) \in \mathbb{R} \times \mathbb{Q}_{p_1,\epsilon}^{n_1-1} \times \mathbb{Q}_{p_2,1}^{n_2-1} \right\},$$

*where $(a, b) := \left( \sqrt{\frac{2n_1-1}{2n}}, \sqrt{\frac{2n_2-1}{2n}} \right)$ and $\epsilon = 1$ or $-1$;*

(iii) *a generalized Lagrangian catenoid*

$$\left\{ (\gamma_1 x, \gamma_2 y) \in \mathbb{Q}_{2p,1}^{2n+1} \,\middle|\, (x, y) \in \mathbb{Q}_{p_1,\epsilon}^{n_1-1} \times \mathbb{Q}_{p_2,1}^{n_2-1}, \quad \tilde{\gamma} = (\gamma_1, \gamma_2) \in \Gamma \right\},$$

*where $\Gamma = \tilde{\gamma}(\mathbb{R})$ is the image of a Legendrian curve $\tilde{\gamma}$ of $\mathbb{Q}_{1-\epsilon,1}^3$ whose projection is a curve $\gamma = \frac{1}{2}(r, x_1, x_2)$ of $\mathbb{Q}_{1-\epsilon,1/4}^2$. The curve $\gamma$, which has curvature $\kappa$ with respect to the unit normal vector $\nu = (\nu_1, \nu_2, \nu_3)$, is solution of the equation*

$$\kappa + \nu_1 \left( \frac{n_1 - 1}{1 - r} - \frac{n_2 - 1}{1 + r} \right) = 0. \tag{5.8}$$

*In the spherical case $\epsilon = 1$, this equation has a countable family of closed solutions and the corresponding solutions are closed, complete minimal Lagragian submanifolds. Moreover, if $p = 0$, i.e. in the case of $\mathbb{CP}^n$, these submanifolds are moreover compact. The other spherical solutions of Equation (5.8) are not properly immersed in $\mathbb{S}^3$ and their images are dense in an open subset of $\mathbb{S}^3$.*

*In the hyperbolic case $\epsilon = -1$, Equation (5.8) has a one-parameter family of solutions which are complete and embedded in $d\mathbb{S}^3$. Therefore the corresponding minimal Lagrangian submanifolds are complete and embedded in $\mathbb{CP}_p^n$.*

## 5.6 Minimal Lagrangian surfaces in the tangent bundle of a Riemannian surface

Consider a surface $\Sigma$ endowed with a positive metric $g$ and an orientation. By Theorem 6 of Chapter 4, this provides a pseudo-Kähler structure $(\Sigma, j, g, \varpi)$, and therefore by the construction made in Section 4.4 of the previous chapter, the tangent bundle $T\Sigma$ enjoys itself a pseudo-Kähler structure $(T\Sigma, \mathbb{J}, \mathbb{G}, \omega)$. The purpose of this section is the local classification of the minimal Lagrangian surfaces of $(T\Sigma, \mathbb{J}, \mathbb{G}, \omega)$. This is based on the rank of projection map $\pi : T\Sigma \to \Sigma$: if $\mathcal{L}$ is a surface of $T\Sigma$ (not necessarily Lagrangian), then the rank of the restriction to $\mathcal{L}$ of the projection $\pi$ can be zero, one or two, and is locally constant. The case of rank zero corresponds to the trivial case of $\mathcal{L}$ being a piece of a vertical fibre.

A simple example of Lagrangian surface of rank one is the *normal bundle* over a curve $\gamma$ of $\Sigma$, i.e. the set of its normal lines to the curve $\gamma$. More precisely, denoting by $\nu(s)$ a unit normal vector to the curve at the point $\gamma(s)$, the normal bundle of $\gamma$ is the image of the immersion $f(s,t) = (\gamma(s), t\nu(s))$. One can slightly generalize the construction by considering affine lines, i.e. adding a translation term to the second factor of the immersion: $f(s,t) = (\gamma(s), a(s)\gamma'(s) + t\nu(s))$, where $a(s)$ is some real-valued function. We shall see in the next section that the image of such an immersion is Lagrangian. We call such a surface an *affine normal bundle over* $\gamma$. Affine normal bundles and their higher dimensional equivalents have been introduced in the flat case in [Harvey, Lawson (1982)], where they were called *degenerate projections*.

**Theorem 32.** *([Anciaux, Guilfoyle, Romon (2009)]) Let $\mathcal{L}$ be a minimal Lagrangian surface of $(T\Sigma, \mathbb{J}, \mathbb{G}, \omega)$. Then*

(i) *either $\mathcal{L}$ is an affine normal bundle over a geodesic on $(\Sigma, g)$, or*

(ii) *$(\Sigma, g)$ is flat.*

**Remark 30.** If $\Sigma$ is flat, its universal covering is $(\mathbb{R}^2, \langle ., .\rangle_0)$, so the universal covering of $T\Sigma$ is $T\mathbb{R}^2 = \mathbb{R}^4$. As explained in Section 4.4.5 of the previous chapter, this is equivalent to $(\mathbb{C}^2, J, \langle ., .\rangle_2, \omega_1)$, whose minimal Lagrangian surfaces have been characterized in Section 5.3 (Theorem 23).

The proof of Theorem 32 results from Propositions 13 and 14 dealing with Lagrangian surfaces of rank one and two respectively, and whose proof are given in the next sections.

### 5.6.1 *Rank one Lagrangian surfaces*

**Proposition 13.** *A rank one Lagrangian surface $\mathcal{L}$ of $(T\Sigma, \mathbb{J}, \mathbb{G}, \omega)$ is an affine normal bundle over a curve $\gamma$ of $\Sigma$. Moreover, $\mathcal{L}$ is minimal if and only if the base curve $\gamma$ is a geodesic of $(\Sigma, g)$.*

*Proof.* A surface $\mathcal{L}$ of $T\Sigma$ with rank one projection may be parametrized locally by

$$f : \quad U \quad \to \quad T\Sigma$$
$$(s,t) \mapsto (\gamma(s), V(s,t)),$$

where $\gamma(s)$ is a regular curve in $\Sigma$ and $V(s,t)$ some tangent vector to $\Sigma$ at the point $\gamma(s)$. Without loss of generality, we may assume that $s$ is the arc length parameter of $\gamma$. Writing $V = a\gamma' + b\nu$ and using the Frénet equation $\gamma'' = \kappa\nu$, where $\kappa$ denotes the curvature of $\gamma$, we compute the first derivatives of the immersion (here and in the following, a letter in subscript denotes partial differentiation with respect to the corresponding variable). Using the direct sum notation:

$$f_s = (\gamma', (a_s - \kappa b)\gamma' + (b_s + \kappa a)\nu),$$
$$f_t = (0, a_t\gamma' + b_t\nu).$$

If the immersion is Lagrangian, the following must vanish:

$$\omega(f_s, f_t) = -g(\gamma', a_t\gamma' + b_t\nu) = -a_t.$$

It follows that $a$ must be a function of $s$. Then we see that for fixed $s$, the map $t \mapsto (\gamma(s), a(s)\gamma'(s) + b(s,t)\nu(s))$ parametrizes a line segment ruled by $\nu(s)$ in $T_{\gamma(s)}\Sigma$, which may be reparametrized by

$$t \mapsto (\gamma(s), a(s)\gamma'(s) + t\nu(s)).$$

We have thus proved the first part of Proposition 13.

We compute easily that $f_s = (\gamma', (a' - \kappa t)\gamma' + a\kappa\nu)$ and $f_t = (0, \nu)$. We also observe that the vector field $f_t$ depends only on the variable $s$, thus it can be extended to a global vector field which is projectable. It follows that we can use Lemma 15 of Chapter 4 (Section 4.4.3) in order to compute the following:

$$D_{f_s} f_t = (0, \nu') = (0, -\kappa\gamma'), \qquad D_{f_t} f_t = (0,0).$$

We recall that by Lemma 16 (Section 5.2), the second fundamental form of $f$ is described by the tri-symmetric tensor

$$T(X, Y, Z) := \mathbb{G}(h(X,Y), \mathbb{J}Z) = \omega(X, D_Y Z),$$

which has four independent components $T_{ijk}$. We calculate, using Lemma 12:

$$T_{112} = \omega(f_s, D_{f_s} f_t) = \omega((\gamma', (a' - \kappa t)\gamma' + a\kappa j\gamma'), (0, -\kappa\gamma')) = \kappa,$$
$$T_{122} = \omega(f_s, D_{f_t} f_t) = 0 \qquad T_{222} = \omega(f_t, D_{f_t} f_t) = 0,$$

(as will become clear in a moment, we do not need the expression of $T_{111}$).

It remains to compute the induced metric, which is given in the coordinates $(s, t)$ by

$$\mathbb{G}_{11} = -2a\kappa, \quad \mathbb{G}_{12} = -1 \quad \text{and} \quad \mathbb{G}_{22} = 0.$$

We are now in a position to get the expression of the mean curvature vector:

$$\mathbb{G}(2\vec{H}, \mathbb{J}f_s) = \frac{T_{111}\mathbb{G}_{22} + T_{122}\mathbb{G}_{11} - 2T_{112}\mathbb{G}_{12}}{\det \mathbb{G}} = -\kappa,$$

and

$$\mathbb{G}(2\vec{H}, \mathbb{J}f_t) = \frac{T_{112}\mathbb{G}_{22} + T_{222}\mathbb{G}_{11} - 2T_{122}\mathbb{G}_{12}}{\det \mathbb{G}} = 0.$$

It follows that

$$2\vec{H} = \kappa \mathbb{J}f_t = (0, \kappa\nu)$$

so that $\mathcal{L}$ is minimal if and only if $\kappa$ vanishes, i.e. $\gamma$ is a geodesic. $\qquad\square$

### 5.6.2   *Rank two Lagrangian surfaces*

**Proposition 14.** *A rank two Lagrangian surface $\mathcal{L}$ of $(T\Sigma, \mathbb{J}, \mathbb{G}, \omega)$ is the graph of the gradient of a real function $u$ on $(\Sigma, g)$:*

$$\mathcal{L} := \{(x, \nabla u(x)), x \in \Sigma\} \subset T\Sigma.$$

*Moreover, if $\mathcal{L}$ is minimal then $g$ is flat (i.e. its Gaussian curvature vanishes).*

*Proof.* A rank two surface is nothing but the graph of a vector field $V(x)$ of $\Sigma$ and thus is the image of the immersion $f(x) = (x, V(x))$. By Theorem 6 of Chapter 2, we know that there exist local coordinates $(s, t)$ on $(\Sigma, g)$ which are isothermic, i.e. $g(s, t) = e^{2r}(ds^2 + dt^2)$, where $r(s, t)$ is a real function. Moreover, we have

$$j\partial_s = \partial_t \quad \text{and} \quad j\partial_t = -\partial_s.$$

Writing $V(s, t) = P(s, t)\partial_s + Q(s, t)\partial_t$, the first derivatives of the immersion are:

$$f_s = (\partial_s, (P_s + Pr_s + Qr_t)\partial_s + (Q_s - Pr_t + Qr_s)\partial_t),$$

$$f_t = (\partial_t, (P_t + Pr_t - Qr_s)\partial_s + (Q_t + Pr_s + Qr_t)\partial_t),$$

so that

$$\omega(f_s, f_t) = g((Q_s - Pr_t + Qr_s)\partial_t, \partial_t) - g(\partial_s, (P_t + Pr_t - Qr_s)\partial_s)$$
$$= e^{2r}(Q_s + 2Qr_s - P_t - 2Pr_t).$$

Thus the Lagrangian condition is equivalent to $(Pe^{2r})_t = (Qe^{2r})_s$, so that there exists locally a real function $u$ on $\Sigma$ such that $Pe^{2r} = u_s$ and $Qe^{2r} = u_t$; in other words, the vector field $V$ is the gradient of $u$, and we have the first part of Proposition 14.

Next a parametrization of $\mathcal{L}$ is

$$f: \Sigma \longrightarrow T\Sigma$$
$$(s,t) \mapsto (x(s,t), e^{-2r}(u_s\partial_s + u_t\partial_t)).$$

We recall from Remark 19 of the previous chapter, the following formulae:

$$\nabla_{\partial_s}\partial_s = r_s\partial_s - r_t\partial_t,$$
$$\nabla_{\partial_t}\partial_s = \nabla_{\partial_s}\partial_t = r_t\partial_s + r_s\partial_t,$$
$$\nabla_{\partial_t}\partial_t = -r_s\partial_s + r_t\partial_t.$$

It follows that

$$f_s = (\partial_s, \nabla_{\partial_s}\nabla u)$$
$$= (\partial_s, e^{-2r}((u_{ss} - 2r_su_s)\partial_s + u_s\nabla_{\partial_s}\partial_s + (u_{st} - 2r_su_t)\partial_t + u_t\nabla_{\partial_s}\partial_t))$$
$$= (\partial_s, e^{-2r}(u_{ss} - r_su_s + r_tu_t)\partial_s + e^{-2r}(u_{st} - r_su_t - r_tu_s)\partial_t).$$

Analogously

$$f_t = (\partial_t, e^{-2r}(u_{st} - r_su_t - r_tu_s)\partial_s + e^{-2r}(u_{tt} - r_tu_t + r_su_s)\partial_t).$$

Denoting for simplicity

$$f_s := (\partial_s, a\partial_s + b\partial_t) \qquad f_t := (\partial_t, b\partial_s + c\partial_t),$$

the induced metric is given by

$$\mathbb{G}_{11} = \omega(\mathbb{J}f_s, f_s) = g(j(a\partial_s + b\partial_t), \partial_s) - g(j\partial_s, a\partial_s + b\partial_t) = -2be^{2r},$$
$$\mathbb{G}_{12} = \omega(\mathbb{J}f_s, f_t) = g(j(a\partial_s + b\partial_t), \partial_t) - g(j\partial_s, b\partial_s + c\partial_t) = (a - c)e^{2r},$$
$$\mathbb{G}_{22} = \omega(\mathbb{J}f_t, f_t) = g(j(b\partial_s + c\partial_t), \partial_t) - g(j\partial_t, b\partial_s + c\partial_t) = 2be^{2r}.$$

Moreover, the vector fields $f_s$ and $f_t$ admit extensions on $T\Sigma$ which are projectable, to that we can use Lemma 15 of Chapter 4 (Section 4.4.3), to get

$$\Pi D_{f_s}f_s = r_s\partial_s - r_t\partial_t,$$
$$KD_{f_s}f_s = (a_s + ar_s + br_t)\partial_s + (b_s - ar_t + br_s)\partial_t + u_se^{-2r}jR(\partial_s, \partial_t)\partial_s,$$
$$\Pi D_{f_s}f_t = r_t\partial_s + r_s\partial_t,$$
$$KD_{f_s}f_t = (b_s + br_s + cr_t)\partial_s + (c_s - br_t + cr_s)\partial_t,$$
$$\qquad + \left(- R(\partial_s, \partial_t)\nabla u + u_se^{-2r}jR(\partial_s, \partial_t)\partial_t + u_te^{-2r}jR(\partial_s, \partial_t)\partial_s\right),$$
$$\Pi D_{f_t}f_t = -r_s\partial_s + r_t\partial_t,$$
$$KD_{f_t}f_t = (b_t + br_t - cr_s)\partial_s + (c_t + br_s + c_t)\partial_t - u_te^{-2r}jR(\partial_t, \partial_s)\partial_t.$$

This allows us to calculate the components of the tensor $T$:

$$
\begin{aligned}
T_{111} &= T(f_s, f_s, f_s) = \omega(f_s, D_{f_s} f_s) \\
&= g(a\partial_s + b\partial_t, r_s\partial_s - r_t\partial_t) \\
&\quad - g(\partial_s, (a_s + ar_s + br_t)\partial_s) - u_s e^{-2r} g(\partial_s, jR(\partial_s, \partial_t)\partial_s) \\
&= e^{2r}(ar_s - br_t - (a_s + ar_s + br_t)) + u_s e^{-2r} g(\partial_t, R(\partial_s, \partial_t)\partial_s) \\
&= e^{2r}(-a_s - 2br_t + u_s K).
\end{aligned}
$$

and similarly[1] with the other coefficients of $T$. Consequently

$$
\begin{aligned}
\mathbb{G}(2\vec{H}, \mathbb{J} f_s) &= \frac{T_{111}\mathbb{G}_{22} + T_{122}\mathbb{G}_{11} - 2T_{112}\mathbb{G}_{12}}{\det \mathbb{G}} \\
&= \frac{1}{2} \frac{2b(T_{111} - T_{122}) - 2(a - c)T_{112}}{-e^{2r}(4b^2 + (a - c)^2)} \\
&= \frac{2b((a - c)_s + 4br_t) + 2(a - c)(-b_s + (a - c)r_t))}{4b^2 + (a - c)^2} \\
&= \frac{(a - c)_s(2b) - (a - c)2b_s}{(2b)^2 + (a - c)^2} + 2r_t \\
&= (\arg(2b + i(a - c))_s + 2r_t.
\end{aligned}
$$

A similar computation yields

$$
\mathbb{G}(2\vec{H}, \mathbb{J} f_t) = (\arg(a - c) + 2ib))_t - 2r_s,
$$

and hence the vanishing of $\vec{H}$ implies

$$
\begin{aligned}
(\arg(c - a + 2ib))_s - 2r_t &= 0, \\
(\arg(c - a + 2ib))_t + 2r_s &= 0.
\end{aligned}
$$

Differentiating the first equation with respect to the variable $t$, and the second equation with respect to the variable $s$ yields $\Delta r = 0$, which implies that $\Sigma$ has vanishing curvature and concludes the proof of Proposition 14.

## 5.7   Exercises

(1) Show that a non-degenerate complex submanifold of a pseudo-Kähler manifold is not only minimal, but austere (see Exercise 2, Chapter 1).

---

[1]Note that coefficients $T_{112}$ and $T_{122}$ can be computed by two different methods, yielding two seemingly different expressions.

(2) Let $f : \mathcal{L} \to \mathbb{Q}^{2n-1}_{2p,\epsilon}$ be a Legendrian immersion and $\gamma \in \mathbb{C}$ a planar, regular curve such that $\gamma \neq 0$. Prove that the map $g$ defined on $I \times \mathcal{L}$ by $g(s,x) = \gamma(s)f(x)$ is a Lagrangian immersion in $\mathbb{C}^n$. Denoting by $\beta_f$ the Lagrangian angle of $f$ and by $\beta_g$ the Legendrian angle of $g$, prove that

$$\beta_f = \beta_g + \arg(\gamma' \gamma^{n-1}).$$

Deduce that $f$ is minimal if and only if $g$ is minimal and $\gamma$ is such that $\arg(\gamma' \gamma^{n-1})$ is constant. This generalizes the construction of Section (5.4.2) (see also [Ros, Urbano (1998)]).

(3) Let $\mathcal{S}$ be a non-degenerate, orientable hypersurface of $\mathbb{Q}^{n+1}_{p,1}$ and let $N(x)$ denote its unit normal field. Set $\epsilon := |N|^2_p$ and

$$\mathcal{L} := \{z = \cos\theta\, x + i\sin\theta\, N(x) | \, x \in \mathcal{S}, \theta \in \mathbb{R}\} \subset \mathbb{Q}^{2n+1}_{2p,1}$$

if $\epsilon = 1$ and

$$\mathcal{L} := \{z = \cosh\theta\, x + i\sinh\theta\, N(x) | \, x \in \mathcal{S}, \theta \in \mathbb{R}\} \subset \mathbb{Q}^{2n+1}_{2p,1}$$

if $\epsilon = -1$. Show that $\mathcal{L}$ is a Legendrian submanifold $\mathbb{Q}^{2n+1}_{2p,1}$. Prove that $\mathcal{L}$ is minimal if and only if $\mathcal{S}$ is austere (see Exercise 2 of Chapter 1). This is a generalization of a construction made in [Borrelli, Gorodski (2004)].

(4) A *cone* of pseudo-Euclidean space $\mathbb{R}^m$ is submanifold $\mathcal{C}$ made of half straight lines passing through the origin. In other words, $\forall (x, \lambda) \in \mathcal{C} \times (0, \infty), \lambda.x \in \mathcal{C}$. Observe that the intersection of a cone with the quadric $\mathbb{Q}^{m-1}_{p,1}$ (sometimes called the *link* of $\mathcal{C}$) is a submanifold of $\mathbb{Q}^{m-1}_{p,1}$. Prove that the link $\mathcal{L} := \mathcal{C} \cap \mathbb{Q}^{2n+1}_{2p,1}$ of a cone $\mathcal{C}$ of $\mathbb{C}^{n+1}$ is Legendrian if and only if $\mathcal{C}$ is Lagrangian with respect to $\omega_p$. Moreover, denoting by $\beta_{\mathcal{C}}$ and $\beta_{\mathcal{L}}$ the Lagrangian angle of $\mathcal{C}$ and the Legendrian angle of $\mathcal{L}$ respectively, we have

$$\beta_{\mathcal{C}}(\lambda x) = \beta_{\mathcal{L}}(x), \ \forall (x, \lambda) \in \mathcal{L} \times (0, \infty).$$

In particular, a Lagrangian cone of $(\mathbb{C}^{n+1}, \langle\langle ., . \rangle\rangle_p)$ is minimal if and only if its link is minimal.

(5) A non-degenerate submanifold $\mathcal{S}$ of a pseudo-Riemannian manifold $(\mathcal{M}, g)$ is said to be *marginally trapped* if its mean curvature vector is null. This thus a generalization of the notion of minimal submanifold. Prove that a marginally trapped hypersurface must be minimal. Hence the notion of marginally trapped submanifold is relevant only in higher co-dimension.

(6) Using the computation of Section 5.4.2, prove that a marginally trapped, $SO(p, n - p)$-equivariant Lagrangian submanifold of $(\mathbb{C}^n, \langle\langle ., . \rangle\rangle_p)$ must be minimal. Consider the Lagrangian immersion $f$ of Exercise 2 and give a condition on $g$ and $\gamma$ equivalent to the fact that $f$ is marginally trapped.

(7) Show that the rank one Lagrangian surfaces of $T\Sigma$ described in Section 5.5.1 are marginally trapped.

# Chapter 6

# Minimizing properties of minimal submanifolds

In the previous chapters, we have described a great variety of minimal submanifolds. On the other hand, recall from our discussion of Chapter 1 that minimality is the first order condition for a submanifold to be a minimizer or a maximizer of the volume. In this last chapter, we shall address the question of whether or not a minimal submanifold is an extremum of the volume. After describing in the first section some classes of minimal submanifolds that do extremize the volume, we shall see in the second section that a necessary condition for this to happen is that the induced metric of both the tangent and normal bundles is definite.

## 6.1 Minimizing submanifolds and calibrations

**Definition 14.** Let $S$ and $S'$ be two submanifolds of a pseudo-Riemannian manifold $(\mathcal{M}, g)$. We shall write $S \sim S'$ if $S$ and $S'$ have the same causal character and $\partial S = \partial S'$.

**Definition 15.** Let $S$ be non-degenerate submanifold of a pseudo-Riemannian manifold $(\mathcal{M}, g)$. Then $S$ is said to be *volume minimizing* (resp. *volume maximizing*) if, for each relatively compact open subset $U$ of $S$,

$$\text{Vol}(U) \leq \text{Vol}(U'), \quad (\text{resp.} \quad \text{Vol}(U) \geq \text{Vol}(U')), \quad \forall U \sim U'.$$

### 6.1.1 *Hypersurfaces in pseudo-Euclidean space*

In this section, we consider the space $\mathbb{R}^{n+1} = \mathbb{R}^n \oplus \mathbb{R}$ endowed with the metric

$$\langle .,. \rangle_{p'} := \langle .,. \rangle_p + \epsilon_{n+1} dx_{n+1}^2$$

and a hypersurface $\mathcal{S}$ of $\mathbb{R}^{n+1} = \mathbb{R}^n \oplus \mathbb{R}$ which is a graph with respect to the coordinate hyperplane $\{x_{n+1} = 0\}$, i.e. $\mathcal{S}$ takes the form

$$\mathcal{S} := \{(x, u(x)) \in \mathbb{R}^{n+1} \,|\, (x_1, ..., x_n) \in U_0\},$$

where $U_0$ is an open subset of $\mathbb{R}^n$ and $u$ is smooth, real function on $U_0$. There is no loss of generality in assuming that $\epsilon_{n+1} = 1$ and we shall do so from now on. We introduce the function $W(x) := \left| |\nabla u(x)|_p^2 + 1 \right|^{1/2}$ and the following vector field of $U_0 \times \mathbb{R}$:

$$X(x, x_{n+1}) := \frac{1}{W(x)} (\nabla u(x), -1).$$

**Lemma 23.** *The restriction of $X$ to $\mathcal{S}$ is a unit normal vector field. Moreover, the formula*

$$nH = -\mathrm{div}X$$

*holds on $\mathcal{S}$. In particular, $\mathcal{S}$ is minimal if and only if $X$ is divergence free (i.e. $\mathrm{div}X = 0$).*

*Proof.* A parametrization of $\mathcal{S}$ being

$$
\begin{aligned}
f : U_0 &\to \mathbb{R}^{n+1} \\
x &\mapsto (x, u(x)),
\end{aligned}
$$

a basis of tangent vector is

$$f_{x_i} = (e_i, u_{x_i}), \forall i, 1 \le i \le n,$$

where $(e_1, ..., e_n)$ denotes the canonical basis of $\mathbb{R}^n$. We claim that $\xi := (\nabla u, -1)$ is normal to $\mathcal{S}$. This follows from the fact that

$$\langle f_{x_i}, \xi \rangle_{p'} = \langle \nabla u, e_i \rangle_p - u_{x_i} = 0,$$

according to the definition of the gradient. Since $|\xi|_{p'}^2 = |\nabla u|_p^2 + 1$, we obtain the first assertion of the lemma: $X := \frac{1}{W}\xi$ is a unit normal vector to $\mathcal{S}$.

We now proceed to calculate the mean curvature of $\mathcal{S}$, starting with its first and second fundamental forms: setting as usual $\epsilon_i := |e_i|_p^2$, we have

$$g_{ij} = \epsilon_i \delta_{ij} + u_{x_i} u_{x_j}.$$

A simple computation shows that

$$g^{ij} = \epsilon_i \delta_{jk} - \frac{\epsilon_j \epsilon_k}{W^2} u_{x_i} u_{x_j}.$$

On the other hand, we have

$$f_{x_i x_j} = (0, u_{x_i x_j}),$$

so the coefficients of the second fundamental form with respect to $N$ are

$$h_{ij} = -\frac{1}{W}u_{x_i x_j}.$$

It follows that

$$nH = -\sum_{i,j=1}^{n} g^{ij}h_{ij}$$

$$= -\frac{1}{W}\sum_{i=1}^{n}\epsilon_i u_{x_i x_i} + \frac{1}{W^3}\sum_{i,j=1}^{n}\epsilon_i\epsilon_j u_{x_i}u_{x_j}u_{x_i x_j}$$

$$= \frac{1}{W^3}\sum_{i,j=1}^{n}\epsilon_i\left(-(\epsilon_j u_{x_j}^2 + 1)u_{x_i x_i} + \epsilon_j u_{x_i}u_{x_j}u_{x_i x_j}\right).$$

On the other hand, taking into account that $X$ does not depend on the variable $x_{n+1}$, we have the following expression for its divergence:

$$divX = \sum_{i=1}^{n}\epsilon_i\langle\nabla_{e_i}X, e_i\rangle_p$$

$$= \sum_{i=1}^{n}\epsilon_i\frac{\partial}{\partial x_i}\left(\frac{u_{x_i}}{W}\right)$$

$$= \sum_{i=1}^{n}\epsilon_i\left(\frac{u_{x_i x_i}}{W} - \frac{u_{x_i}W_{x_i}}{W^2}\right)$$

$$= \frac{1}{W^3}\sum_{i,j=1}^{n}\epsilon_i\left((\epsilon_j u_{x_j}^2 + 1)u_{x_i x_i} - \epsilon_j u_{x_i}u_{x_j}u_{x_i x_j}\right).$$

Therefore the equation $nH = -divX$ holds as claimed. $\qquad\square$

**Remark 31.** From these calculations we deduce that the PDE satisfied by a minimal graph is:

$$\sum_{i,j=1}^{n}\epsilon_i\left((\epsilon_j u_{x_j}^2 + 1)u_{x_i x_i} - \epsilon_j u_{x_i}u_{x_j}u_{x_i x_j}\right) = 0.$$

This generalizes both Equations (2.2) and (2.6) of Chapter 2.

**Theorem 33.** *Let $S$ be a non-degenerate minimal graph of pseudo-Euclidean space $(\mathbb{R}^{n+1}, \langle., .\rangle_{p'})$ with definite induced metric (so in particular $\langle., .\rangle_{p'}$ is Riemannian or Lorentzian). Then $S$ is volume minimizing if $\langle., .\rangle_{p'}$ is Riemannian and volume maximizing if $\langle., .\rangle_{p'}$ is Lorentzian.*

**Remark 32.** In the Lorentzian case, a hyperplane with definite metric is never vertical. It follows that a connected hypersurface with definite metric is always a graph. Therefore the result above can be stated as follows: a spacelike minimal hypersurface of Minkowski space is volume maximizing. This is why such surfaces are usually called *maximal*. Locally, a hypersurface is always a graph on its tangent plane. Therefore, the result above implies that in the Euclidean case a minimal hypersurface is always volume minimizing for *local* variations.

*Proof of Theorem 33.* There is no loss of generality in assuming that $S$ is a graph with respect to the hyperplane $\{x_{n+1} = 0\}$. Let $U$ be a open subset of $S$ and $U' \sim U$. We use the divergence theorem (Theorem 2, Chapter 1), with the vector field $X$, denoting by $T$ the open subset of $\mathbb{R}^{n+1}$ whose boundary is $\partial T = U - U'$. Hence by Lemma 23, we have

$$0 = \int_T div X \, dV = \epsilon \int_U \langle X, N \rangle_{p'} dV - \epsilon \int_{U'} \langle X, N \rangle_{p'} dV,$$

where $N$ denotes the outward unit normal vector of $\partial T$ and $\epsilon := |N|_{p'}^2$ (since $U$ and $U'$ have the same causal character, $\epsilon$ is constant on $U - U'$). Moreover, since $X = N$ on $S$ and does not depend on the variable $x_{n+1}$, $|X|_{p'} = \epsilon$ along $U'$ as well.

If the metric $\langle ., . \rangle_{p'}$ is definite, $X^T$ and $N$ have the same causal character, so there exists, locally, an $\mathbb{R}/2\pi\mathbb{Z}$-valued map $\theta$ and a unit tangent vector field $e_0$ on $U'$ such that

$$X = \cos\theta N + \sin\theta \, e_0.$$

It follows that

$$\mathrm{Vol}(U) = \int_U dV = \int_U \epsilon \langle X, N \rangle_{p'} dV$$

$$= \int_{U'} \epsilon \langle X, N \rangle_{p'} dV = \int_{U'} \cos\theta \, dV \leq \mathrm{Vol}(U'),$$

i.e. $S$ is volume minimizing.

If the metric $\langle ., . \rangle_{p'}$ is indefinite, $X^T$ and $N$ have opposite causal characters, so there exists, locally, a real map $\theta$ and a unit tangent vector field $e_0$ on $S'$ such that

$$X = \cosh\theta N + \sinh\theta \, e_0,$$

and we have

$$\mathrm{Vol}(U) = \int_U dV = \int_U \epsilon \langle X, N \rangle_{p'} dV$$

$$= \int_{U'} \epsilon \langle X, N \rangle_{p'} dV = \int_{U'} \cosh\theta \, dV \geq \mathrm{Vol}(U'),$$

hence $S$ is volume maximizing. □

**Remark 33.** In the Riemannian case $p' = 0$, we could have used as well the Cauchy-Schwartz inequality $\langle X, N \rangle_0 \leq |X|_0 |N|_0 \leq 1$ to get the conclusion.

### 6.1.2 Complex submanifolds in pseudo-Kähler manifolds

The following result is the generalization to the pseudo-Riemannian case of the well known fact that complex submanifolds of Kähler manifolds are calibrated (see [Harvey, Lawson (1982)],[Harvey (1990)]). The fundamental inequality appearing in the proof is called *Wirtinger's inequality*.

**Theorem 34.** *Let $S$ be a complex submanifold of a pseudo-Kähler manifold $(\mathcal{M}, J, g, \omega)$ such that the induced metric on both tangent and normal bundles is definite. Then if $g$ is definite (resp. indefinite) $S$ is volume minimizing (resp. maximizing).*

*Proof.* Assume without loss of generality that the induced metric on $TS$ is positive. Let $U$ be a relatively compact open subset of $S$ and $U' \sim U$. In particular, $U'$ is non-degenerate and the induced metric on $TU'$ is positive. At a given point $x$ of $U'$, introduce the linear map

$$J^\top : T_x U' \to T_x U'$$
$$X \mapsto (JX)^\top.$$

It is skew-symmetric with respect to $g$, since

$$g((JX)^\top, Y) = g(JX, Y) = \omega(X, Y) = -\omega(Y, X) = -g((JY)^\top, X).$$

Denoting by $2k$ the dimension of $U$, there exists an orthonormal basis $(e_1, ..., e_{2k})$ such that the matrix of $J^\top$ takes the following form

$$\begin{pmatrix} 0 & \lambda_1 & ... & 0 & 0 \\ -\lambda_1 & 0 & ... & 0 & 0 \\ \vdots & & & & \vdots \\ 0 & 0 & ... & 0 & \lambda_k \\ 0 & 0 & ... & -\lambda_k & 0 \end{pmatrix},$$

where the coefficients $\lambda_i, 1 \leq i \leq k$, are positive. Hence $(Je_{2i-1})^\top = \lambda_i e_{2i}$ and

$$1 = g(e_{2i-1}, e_{2i-1}) = g(Je_{2i-1}, Je_{2i-1})$$
$$= g((Je_{2i-1})^\top, (Je_{2i-1})^\top) + g((Je_{2i-1})^\perp, (Je_{2i-1})^\perp)$$
$$= \lambda_i^2 + g((Je_{2i-1})^\perp, (Je_{2i-1})^\perp)$$

If the metric on the normal bundle of $U'$ is negative, $g((Je_{2i-1})^\perp,$ $(Je_{2i-1})^\perp) \leq 0$ is negative, so $\lambda_i \geq 1$. We deduce that

$$\omega^k(e_1, ..., e_{2k}) = \Pi_{i=1}^k \omega(e_{2i-1}, e_{2i}) = \Pi_{i=1}^k \lambda_i \geq 1 = dV(e_1, ..., e_{2k}).$$

Since $\omega^k$ and $dV$ have maximal degree $2k$, we get the (generalized) Wirtinger inequality: $\omega^k \geq dV$. Moreover, equality occurs if and only if all coefficients $\lambda_i$s are equal to 1. By the computation above this implies that $g((Je_{2i-1})^\perp, (Je_{2i-1})^\perp)$ vanishes, i.e. $Je_{2i-1} \in T_xU'$. Since this holds $\forall i, 1 \leq i \leq k$, the tangent plane $T_xU'$ is complex.

Finally, using the fact that $d\omega^k = kd\omega \wedge \omega^{k-1} = 0$ and Stokes theorem, we obtain

$$\mathrm{Vol}(U') = \int_{U'} dV = \int_{U'} \omega^k = \int_S \omega^k \geq \int_S dV = \mathrm{Vol}(\mathcal{S}),$$

with equality if and only if $\mathcal{S}'$ is a complex submanifold. In the language of [Harvey, Lawson (1982)], $\omega^k$ is a *calibration*, whose calibrated submanifolds are the $2k$-dimensional complex submanifolds with definite tangent and normal bundles.

We conclude that a complex submanifold such that the induced metric on the tangent (resp. normal) bundle is positive (resp. negative) is volume maximizing. If the normal bundle is positive instead of negative, all the inequalities are reversed (in particular the $\lambda_i$s are less than 1) and we deduce that the complex submanifolds of pseudo-Kähler manifolds with definite metric are volume minimizing. □

### 6.1.3  *Minimal Lagrangian submanifolds in complex pseudo-Euclidean space*

**Theorem 35 ([Harvey, Lawson (1982)]).** *Let $\mathcal{L}$ be a minimal Lagrangian submanifold of $(\mathbb{C}^n, \langle\langle ., . \rangle\rangle_0)$. Then $\mathcal{L}$ is volume minimizing.*

*Proof.* We have seen in Chapter 5, Theorem 24, that the Lagrangian angle of a minimal Lagrangian submanifold $\mathcal{L}$ of $(\mathbb{C}^n, \langle\langle ., . \rangle\rangle_0)$ is constant. Let us denote by $\beta_0$ this constant and introduce the $n$-form on $\mathbb{C}^n$ defined by

$$\Theta := \mathrm{Re}\left(e^{-i\beta_0}\Omega\right) = \mathrm{Re}\left(e^{-i\beta_0}(dz_1 \wedge ... \wedge dz_n)\right).$$

We claim that $\Theta$ is a calibration, i.e. it is closed and $\Theta \leq dV$, with equality if and only if $\Theta$ is evaluated on a Lagrangian subspace with Lagrangian angle $\beta_0$. To see it is sufficient to prove that $\Theta(f_1, ..., f_n) \leq 1$, where $(f_1, ..., f_n)$ is an orthonormal family, with equality if and only if

$\beta_0(f_1, ..., f_n) = \beta_0$. For this purpose we introduce the complex-valued matrix $M := [\langle\langle f_j, e_k \rangle\rangle_0]_{1 \leq j,k \leq n}$, where $(e_1, ..., e_n)$ is the canonical basis of $\mathbb{C}^n$. In particular, $\operatorname{Re} M = [\langle f_j, e_k \rangle_0]_{1 \leq j,k \leq n}$ and $\operatorname{Im} M = [\langle f_j, Je_k \rangle_0]_{1 \leq j,k \leq n}$. From the fact that

$$f_j = \sum_{k=1}^{n} (\langle f_j, e_k \rangle_0 e_k + \langle f_j, Je_k \rangle_0 Je_k) = \sum_{k=1}^{n} \langle\langle f_j, e_k \rangle\rangle_0 e_k$$

we have

$$(\Theta(f_1, ..., f_n))^2 \leq |\Omega(f_1, ..., f_n)|^2$$
$$= |\det_{\mathbb{C}} M|^2$$
$$= \left| \det_{\mathbb{R}} \begin{pmatrix} \operatorname{Re} M & \operatorname{Im} M \\ -\operatorname{Im} M & \operatorname{Re} M \end{pmatrix} \right|$$
$$\leq \Pi_{j=1}^{n} |f_j|_0 |Jf_j|_0 = 1.$$

where we have used Hadamard's inequality for matrices in the last inequality (see [Harvey (1990)]). Moreover, equality occurs if and only

$$\Theta(f_1, ..., f_n) = \Omega(f_1, ..., f_n) \quad \text{and} \quad \left| \det_{\mathbb{R}} \begin{pmatrix} \operatorname{Re} M & \operatorname{Im} M \\ -\operatorname{Im} M & \operatorname{Re} M \end{pmatrix} \right| = 1.$$

The second equality holds if and only if the family $(f_1, ..., f_n, Jf_1, ..., Jf_n)$ is orthonormal, which is equivalent to the fact that the family $(f_1, ..., f_n)$ spans a Lagrangian subspace. The first equality is equivalent to the fact that $\Theta(f_1, ..., f_n) = \operatorname{Re} \Omega(f_1, ..., f_n)$, i.e. $\arg \Omega(f_1, ..., f_n) = \beta_0$, so finally, $\Theta(f_1, ..., f_n) = 1$ if and only if $Span(f_1, ..., f_n)$ is a Lagrangian subspace with Lagrangian angle $\beta_0$.

To conclude the proof, observe that since $\Theta$ has constant coefficients, it is closed, so we may use Stokes theorem as follows: given a relatively compact open subset $U$ of $\mathcal{L}$ and $U' \sim U$, we have

$$\operatorname{Vol}(U) = \int_U dV = \int_U \Theta = \int_{U'} \Theta \leq \int_{U'} dV = \operatorname{Vol}(U'). \qquad \square$$

**Remark 34.** This result no longer holds true if the metric is indefinite, as proves the following counter-example[1]: consider the case of $(\mathbb{C}^2, \langle\langle ., . \rangle\rangle_1)$ and $X_1 = (1, ic)$, $X_2 = (ic^{-1}, 1)$, where $c$ is a non-vanishing, real constant. Hence $\Omega(X_1, X_2) = 2$, while $dV(X_1, X_2) = c^{-2}(c^2 - 1)$, which can be set arbitrarily close to zero by letting $c$ tend to 1. Observe that the span of $(X_1, X_2)$ is never Lagrangian, unless $c = 1$, in which case it becomes degenerate. The next theorem says that the inequality $|\Omega| \leq dV$ holds in the indefinite case if we restrict ourselves to Lagrangian non-degenerate subspaces.

---

[1]This is also a counter-example showing that Hadamard inequality no longer holds for an indefinite metric.

**Remark 35.** Theorem 35 may be extended to a certain class of Kähler manifolds, called *Calabi-Yau* manifolds. Roughly speaking, a Calabi-Yau manifold is a Kähler manifold where it is possible to define globally a holomorphic $n$-form playing the rôle of $\Omega$. Minimal Lagrangian submanifolds of Calabi-Yau manifolds are usually called *special Lagrangian* manifolds.

**Theorem 36.** *Let $\mathcal{L}$ be a minimal Lagrangian submanifold of $(\mathbb{C}^n, \langle\langle .,.\rangle\rangle_p)$. Then $\mathcal{L}$ is volume minimizing with respect to Lagrangian variations, i.e. for each relatively compact open subset $U$ of $\mathcal{L}$,*

$$Vol(U) \leq Vol(U'), \quad \forall U' \sim U, \ U' \ Lagrangian.$$

*Proof.* As in the Theorem 35, the proof is based on the fact that the Lagrangian angle of $\mathcal{L}$ is equal to a constant $\beta_0$ and the analysis of the behaviour of the $n$-form $\Theta$. We claim that if $X_1, ..., X_n$ are $n$ vectors spanning a non-degenerate Lagrangian subspace, then $\Theta(X_1, ..., X_n) \leq dV(X_1, ..., X_n)$, with equality if and only if $\beta(X_1, ..., X_n) = \beta_0$. We have $\Omega(X_1, ..., X_n) = \det_{\mathbb{C}} M$, with $M = [\langle\langle X_j, e_k\rangle\rangle_p]_{1 \leq j,k \leq n}$. On the other hand, by the Lagrangian assumption, we have

$$\langle X_j, X_k\rangle_p = \langle\langle X_j, X_k\rangle\rangle_p = \sum_{l=1}^{n} \epsilon_l \langle\langle X_j, e_l\rangle\rangle_p \langle\langle X_k, e_l\rangle\rangle_p.$$

Therefore

$$dV(X_1, ..., X_n) = |\det([\langle X_j, X_k\rangle_p]_{1 \leq j,k \leq n})|^{1/2}$$
$$= |\det(M I_p M^*)|^{1/2} = \left|\det_{\mathbb{C}} M\right|,$$

where $M^*$ denotes the Hermitian adjoint of $M$ and $I_p$ stands for the diagonal matrix $diag(\epsilon_1, ..., \epsilon_n)$. It follows that

$$\Theta(X_1, ..., X_n) \leq |\Omega(X_1, ..., X_n)| = \left|\det_{\mathbb{C}} M\right| = dV(X_1, ..., X_n),$$

and of course equality holds if and only if $\beta(X_1, ..., X_n) = \arg(\Omega(X_1, ..., X_n)) = \beta_0$. The conclusion of the proof follows exactly the lines of that of Theorem 35. $\square$

## 6.2   Non-minimizing submanifolds

We may observe that all the extremizing submanifolds discussed in the previous section (with the exception of the rather special case of Theorem 36) share a common property: the induced metric of both their tangent and normal bundles is definite. The next theorem shows that this property is actually a necessary condition:

**Theorem 37.** *A minimal submanifold of a pseudo-Riemannian manifold such that its tangent or its normal bundle is indefinite is unstable.*

**Remark 36.** This result is a slight generalization of Theorem A of [Dong (2009)].

**Remark 37.** The condition that the tangent or the normal bundle of a minimal submanifold is indefinite is necessary but not sufficient. For example, many minimal submanifolds in Riemannian manifolds are unstable (e.g. the totally geodesic spheres of $(\mathbb{S}^{n+1}, \langle ., . \rangle_0)$).

We have seen in Chapter 5 (Lemma 16) that the induced metric of a Lagrangian submanifold of an indefinite pseudo-Kähler manifold is itself indefinite. Hence Theorem 37 implies in particular:

**Corollary 9.** *A minimal Lagrangian submanifold of a pseudo-Kähler manifold with indefinite metric is unstable.*

*Proof of Theorem 37.* Choose an open subset $U$ of $S$ which is small enough to enjoy a system of coordinates $(x_1, ..., x_n)$ and a normal vector field $\xi$ whose compact support is contained in $U$ and which does not change causal character (it is either positive or negative). Consider the family of variations $S_t$ induced by the normal vector fields $X := \sin(kx_1)\xi$, $k \in \mathbb{N}$; we have seen in Chapter 1, Section 1.3 that the second variation formula is

$$\frac{d^2}{dt^2}\text{Vol}(S_t)\bigg|_{t=0} = \int_U \left( g(\nabla^\perp X, \nabla^\perp X) - g(A_X, A_X) + g(R^\perp(X), X) \right) dV.$$

Observe that

$$-g(A_X, A_X) + g(R^\perp(X), X) = \sin^2(kx_1)\left( -g(A_\xi, A_\xi) + g\left(R^\perp(\xi), \xi\right) \right),$$

so there is a constant $C$ independent of $k$ such that

$$\left| \int_S \left( -g(A_X, A_X) + g(R^\perp(X), X) \right) dV \right| \le C.$$

On the other hand,

$$g(\nabla^\perp X, \nabla^\perp X) = \sum_{i,j=1}^n g^{ij} g(\nabla^\perp_{\partial_{x_i}} X, \nabla^\perp_{\partial_{x_i}} X)$$

$$= g^{11} \left( k^2 \cos^2(kx_1) g(\xi, \xi) + 2k \cos(kx_1) \sin(k\xi_1) g(\xi, \nabla^\perp_{\partial_{x_1}} \xi) \right.$$

$$\left. + \sin^2(kx_1) g(\nabla^\perp_{\partial_{x_1}} \xi, \nabla^\perp_{\partial_{x_1}} \xi) \right)$$

$$+ 2\sin(k\xi_1) \sum_{i=2}^n g^{1i} \left( k \cos(kx_1) g(\xi, \nabla^\perp_{\partial_{x_i}} \xi) \right.$$

$$\left. + \sin(k\xi_1) g(\nabla^\perp_{\partial_{x_1}} \xi, \nabla^\perp_{\partial_{x_i}} \xi) \right)$$

$$+ \sin^2(kx_1) \sum_{i,j=2}^n g^{ij} g(\nabla^\perp_{\partial_{x_i}} \xi, \nabla^\perp_{\partial_{x_j}} \xi).$$

It follows that

$$\left| \frac{d^2}{dt^2} \mathrm{Vol}(\mathcal{S}_t) \big|_{t=0} - k^2 \int_U g^{11} \cos^2(kx_1) g(\xi, \xi) dV \right| \leq kC_1 + C_2,$$

where $C_1$ and $C_2$ are two real constants not depending on $k$. Moreover, since $U$ is compact and $g^{11} g(\xi, \xi)$ is smooth on $U$,

$$\lim_{k \to \infty} \int_U g^{11} \cos^2(kx_1) g(\xi, \xi) dV = \frac{1}{2} \int_U g^{11} g(\xi, \xi) dV.$$

Therefore, for $k$ large enough, the second variation has the sign of $g^{11} g(\xi, \xi)$ (which is constant on $U$). If the tangent bundle of $\mathcal{S}$ has indefinite metric, by re-labeling the coordinates, we may choose $x_1$ in such a way that $\partial_{x_1}$ is either a positive or a negative vector field, so that $g^{11}$ is either positive or negative. Analogously, if the normal bundle has indefinite metric, we can choose $\xi$ to be either positive and negative. In both cases, we can let $g^{11} g(\xi, \xi)$ to be positive or negative, which proves the claim. $\square$

# Bibliography

D. Alekseevsky and H. Baum, Eds (2008), *Recent developments in pseudo-Riemannian geometry*, EMS.

H. Alencar (1993), *Minimal Hypersurfaces of $\mathbb{R}^{2m}$ invariant by $SO(m) \times SO(m)$*, Trans. of the Amer. Math. Soc., **337**, no. 1, pp. 129–141.

H. Alencar, A. Barros, O. Palmas, J. Reyes and W. Santos (2005), $O(m) \times O(n)$-*invariant minimal hypersurfaces in $\mathbb{R}^{m+n}$*, Ann. Global Anal. Geom. **27**, no. 2, pp. 179–199.

H. Anciaux (2006), *Legendrian submanifolds foliated by $(n-1)$-spheres in $\mathbb{S}^{2n+1}$*, Mat. Cont. **30**, pp. 41–61.

H. Anciaux, I. Castro and P. Romon (2006), *Lagrangian submanifolds foliated by $(n-1)$-spheres in $\mathbb{R}^{2n}$*, Acta Math. Sinica (English Series) **22**, no. 4, pp. 1197–1214.

H. Anciaux, B. Guilfoyle and P. Romon (2009), *Minimal Lagrangian surfaces in the tangent bundle of a Riemannian surface*, arXiv:0807.1387.

A.I. Bobenko, P. Schröder, J.M. Sullivan and G.M. Ziegler, Eds. (2008), *Discrete differential geometry*, Oberwolfach Seminars, **38**, Birkhäuser.

E. Boeckx and L. Vanhecke (2000), *Geometry of Riemannian manifolds and their unit tangent sphere bundles*, Publ. Math. Debrecen **57**, no. 3-4, pp. 509–533.

E. Bombieri, E. De Giorgi and E. Giusti (1969), *Minimal cones and the Bernstein problem*, Invent. Math. **7**, pp. 243–268.

V. Borrelli and C. Gorodski (2004), *Minimal Legendrian submanifolds of $\mathbb{S}^{2n-1}$ and absolutely area-minimizing cones*, Differential Geom. and its Appl. **21**, no. 3, pp. 337–347.

F. Brito and M.-L. Leite (1990), *A Remark on rotational hypersurfaces in $\mathbb{S}^n$*, Bull. of the Belgian Math. Soc. **42**, pp. 303–318.

E. Calabi (1968), *Examples of Bernstein problems for some nonlinear equations*, Proc. Sympos. Pure Appl. Math. **15**, pp. 223–230.

I. Castro, C. R. Montealegre and F. Urbano (1999), *Minimal Lagrangian submanifolds in the complex hyperbolic space*, Illinois J. Math. **46**, no. 3, pp. 695–721.

I. Castro and F. Urbano (1999), *On a Minimal Lagrangian Submanifold of $\mathbb{C}^n$ Foliated by Spheres*, Mich. Math. J. **46**, pp. 71–82.

I. Castro and F. Urbano (2004), *On a new construction of special Lagrangian immersions in complex Euclidean space,*   Quarter. J. Math. **55**, pp. 253–266.

I. Castro, H. Li and F. Urbano (2006), *Hamiltonian-minimal Lagrangian submanifolds in complex space forms,* Pacific J. Math. **227**, no 1, pp. 43–63.

B.-Y. Chen (1973), *Geometry of Submanifolds,* Dekker, New York.

B.-Y. Chen (2003), *Complex extensors and Lagrangian submanifolds in indefinite complex Euclidean spaces,* Bull. of the Inst. of Math. Acad. Sinica **31**, No. 3.

B.-Y. Chen (2009), *Non linear Klein-Gordon equations and Lorentzian minimal surfaces in Lorentzian complex space forms,* Taiwanese J. of Math. **13**, no. 1, pp. 1–24.

S.-Y. Cheng and S.-T. Yau (1976), *Maximal spacelike hypersurface in Lorentz-Minkowsi space,* Ann. of Math. (2) **104**, pp. 407–419.

M. do Carmo (1976), *Differential geometry of curves and surfaces,* Translated from the Portuguese. Prentice-Hall, Inc., Englewood Cliffs, N.J.

M. do Carmo (1992), *Riemannian geometry,* Birkhäuser.

P. Dombrowski (1962), *On the geometry of the tangent bundle,* J. Reine Angew. Math. **210** pp. 73–88.

Y. Dong (2009), *On Indefinite Special Lagrangian Submanifolds in Indefinite Complex Euclidean Spaces,* J. of Geom. and Physics **59** pp. 710–726.

I. Fernández and F. Lopez (2010), *On the uniqueness of the helicoid and Enneper's surface in the Lorentz-Minkowski space* $\mathbb{R}_1^3$, to appear in Trans. Amer. Math. Soc.

S. Furuya (1971), *On periods of periodic solutions of certain non-linear differential equation,* US-Japan Smeinar of Ordinary Diff. and Funct. Equations, Springer-Verlag pp. 320–323.

B. Guilfoyle and W. Klingenberg (2005), *An indefinite Kähler metric on the space of oriented lines,* J. London Math. Soc. **72** pp. 497–509.

B. Guilfoyle and W. Klingenberg (2008), *On area-stationary surfaces in certain neutral Kähler 4-manifolds,* Beiträge Algebra Geom. **49**, no. 2, pp. 481–490.

R. Harvey and H. B. Lawson (1982), *Calibrated geometries,* Acta Math. **148**, pp. 47–157.

R. Harvey (1990), *Spinors and calibrations,* Academic Press.

M. Haskins and N. Kapouleas (2010), *Twisted products and* $SO(p) \times SO(q)$-*invariant special Lagrangian cones,* arXiv:1005.1419.

F. Hélein and P. Romon (2002), *Hamiltonian stationary Lagrangian surfaces in Hermitian symmetric spaces,* in Differential geometry and integrable systems (Tokyo, 2000), M. Guest, R. Miyaoka and Y. Ohnita Eds, Contemp. Math., **308**, Amer. Math. Soc., Providence, RI, 2002, pp. 161–178.

L. Hörmander (1990), *An introduction to complex analysis in several variables,* third edition, North-Holland Publishing Co., Amsterdam.

D. Joyce (2001), *Constructing special Lagrangian m-folds in* $\mathbb{C}^m$ *by evolving quadrics,* Math. Annalen **320** pp. 757–797.

D. Joyce, Y.-I. Lee and M.-P. Tsui (2010), *Self-similar solutions and translating solitons for Lagrangian mean curvature flow,* J. of Diff. Geom. **84**, no. 1, pp. 127–161.

O. Kobayashi (1983), *Maximal surfaces in the 3-dimensional Minkowski space $L^3$*, Tokyo J. Math. **6**, no. 2, pp. 297–309.

O. Kowalski (1971), *Curvature of the induced Riemannian metric on the tangent bundle of a Riemannian manifold*, J. Reine Angew. Math. **250**, pp. 124–129.

M. Kriele (1999), *Spacetime, Foundations of General Relativity and Differential Geometry*, Springer.

W. Kühnel (2000), *Differential Geometry, Curves – Surfaces – Manifolds*, Student Mathematical Library, **16**, AMS.

J. Lafontaine (1994), *Some relevant Riemannian geometry*, in *Holomorphic curves in symplectic geometry*, J. Lafontaine and M. Audin Eds., Birkhäuser.

B. Lawson (1977), *Lectures on minimal submanifolds* Vol. 1, Monografias de Matemática, 14, IMPA, Rio de Janeiro.

Y.-I. Lee and M.-T. Wang (2009), *Hamiltonian Stationary Shrinkers and Expanders for Lagrangian Mean Curvature Flows*, J. Differential Geom. **83**, no. 1, pp. 27–42.

M. Magid (1989), *The Bernstein problem for timelike surfaces*, Yokohama Math. J. **37**, pp. 125-137.

M. Magid (1991), *Timelike surfaces in Lorentz 3-space with prescribed mean curvature and Gauss map*, Hokkaido Math. J. **19**, pp. 447–464.

W. H. Meeks III and H. Rosenberg (2005), *The uniqueness of the helicoid*, Ann. Math., **161**, pp. 727-758.

A. Moroianu (2007), *Lectures on Kähler Geometry*, London Mathematical Society Student Texts 69, Cambridge University Press, Cambridge.

A. Niang (2003), *Surfaces minimales réglées dans l'espace de Minkowski ou euclidien orienté de dimension 3*, Afrika Mat. **15**, no. 3, pp. 117–127.

A. Newlander and L. Nirenberg (1957), *Complex analytic coordinates in almost complex manifolds*, Annals of Math. (2) **65**, pp. 391-404.

B. Nelli and M. Soret (2005), *The state of the art of Bernstein's problem*, Math. Cont. **29**, pp. 69–77.

J. Nitsche (1989), *Lectures on Minimal Surfaces*, Cambridge Univ. Press.

B. O'Neill (1983), *Semi-Riemannian Geometry with Applications to Relativity*, Academic Press, New York, London.

T. Otsuki (1970), *Minimal Hypersurfaces in a Riemannian Manifold of Constant Curvature*, Amer. J. of Math., **92**, no. 1, pp. 145–173.

R. Osserman (1969), *A survey of minimal surfaces*, Van Nostrand, New York.

F. Palomo and A. Romero (2006), *Certain Actual Topics on Modern Lorentzian Geometry* in Handbook of Differential Geometry, Vol. 2, F. Dillen and L. Verstraelen Eds, Elsevier Science, pp. 513–546.

U. Pinkall (1985), *Hopf tori in $\mathbb{S}^3$*, Invent. Math. **81**, no 2, pp. 379–386.

A. Ros and F. Urbano (1998), *Lagrangian submanifolds of $\mathbb{C}^n$ with conformal Maslov form and the Whitney sphere*, J. Math. Soc. Japan **50**, pp. 203–226.

U. Simon (2000), *Affine differential geometry*, in Handbook of Differential Geometry, Vol. 1, F. Dillen and L. Verstraelen Eds, Elsevier Science, pp. 905–961.

M. Spivak (1965), *Calculus on manifolds, A modern approach to classical Theorems of Advanced Calculus*, Addison-Wesley.

M. Spivak (1979), *A comprehensive introduction to differential geometry*, Publish or Perish, Inc.

Y. Xin (2003), *Minimal submanifolds and related topics*, Nankai tracts in mathematics, Vol. 8, World Scientific.

I. van de Woestijne (1990), *Minimal surfaces of the 3-dimensional Minkowski space*, in Geometry and Topology of submanifolds II., M. Boyom, J.-M. Morvan and L.Verstraelen Eds, World Scientific, Singapore. pp. 344–369.

T. Weinstein (1996), *An introduction to Lorentz surfaces*, de Gruyter Expositions in Mathematics, 22 Walter de Gruyter & Co., Berlin.

# Index